EMPOWERING THE
Defense Acquisition Workforce
TO IMPROVE MISSION OUTCOMES USING DATA SCIENCE

Committee on Improving Defense Acquisition
Workforce Capability in Data Use

Board on Mathematical Sciences and Analytics
Committee on Applied and Theoretical Statistics
Air Force Studies Board
Computer Science and Telecommunications Board
Division on Engineering and Physical Sciences

Board on Higher Education and Workforce
Policy and Global Affairs

Committee on National Statistics
Division of Behavioral and Social Sciences and Education

A Consensus Study Report of
The National Academies of
SCIENCES · ENGINEERING · MEDICINE

THE NATIONAL ACADEMIES PRESS
Washington, DC
www.nap.edu

THE NATIONAL ACADEMIES PRESS   500 Fifth Street, NW   Washington, DC 20001

This activity was supported by contract FA8650-19-F-9301 between the National Academy of Sciences and the Department of Defense. Any opinions, findings, conclusions, or recommendations expressed in this publication do not necessarily reflect the views of any organization or agency that provided support for the project.

International Standard Book Number-13: 978-0-309-68493-4
International Standard Book Number-10: 0-309-68493-5
Digital Object Identifier: https://doi.org/10.17226/25979

This publication is available from the National Academies Press, 500 Fifth Street, NW, Keck 360, Washington, DC 20001; (800) 624-6242 or (202) 334-3313; http://www.nap.edu.

Copyright 2021 by the National Academy of Sciences. All rights reserved.

Printed in the United States of America

Suggested citation: National Academies of Sciences, Engineering, and Medicine. 2021. *Empowering the Defense Acquisition Workforce to Improve Mission Outcomes Using Data Science*. Washington, DC: The National Academies Press. https://doi.org/10.17226/25979.

The National Academies of
SCIENCES • ENGINEERING • MEDICINE

The **National Academy of Sciences** was established in 1863 by an Act of Congress, signed by President Lincoln, as a private, nongovernmental institution to advise the nation on issues related to science and technology. Members are elected by their peers for outstanding contributions to research. Dr. Marcia McNutt is president.

The **National Academy of Engineering** was established in 1964 under the charter of the National Academy of Sciences to bring the practices of engineering to advising the nation. Members are elected by their peers for extraordinary contributions to engineering. Dr. John L. Anderson is president.

The **National Academy of Medicine** (formerly the Institute of Medicine) was established in 1970 under the charter of the National Academy of Sciences to advise the nation on medical and health issues. Members are elected by their peers for distinguished contributions to medicine and health. Dr. Victor J. Dzau is president.

The three Academies work together as the **National Academies of Sciences, Engineering, and Medicine** to provide independent, objective analysis and advice to the nation and conduct other activities to solve complex problems and inform public policy decisions. The National Academies also encourage education and research, recognize outstanding contributions to knowledge, and increase public understanding in matters of science, engineering, and medicine.

Learn more about the National Academies of Sciences, Engineering, and Medicine at **www.nationalacademies.org**.

The National Academies of
SCIENCES · ENGINEERING · MEDICINE

Consensus Study Reports published by the National Academies of Sciences, Engineering, and Medicine document the evidence-based consensus on the study's statement of task by an authoring committee of experts. Reports typically include findings, conclusions, and recommendations based on information gathered by the committee and the committee's deliberations. Each report has been subjected to a rigorous and independent peer-review process and it represents the position of the National Academies on the statement of task.

Proceedings published by the National Academies of Sciences, Engineering, and Medicine chronicle the presentations and discussions at a workshop, symposium, or other event convened by the National Academies. The statements and opinions contained in proceedings are those of the participants and are not endorsed by other participants, the planning committee, or the National Academies.

For information about other products and activities of the National Academies, please visit www.nationalacademies.org/about/whatwedo.

COMMITTEE ON IMPROVING DEFENSE ACQUISITION WORKFORCE CAPABILITY IN DATA USE

WENDY M. MASIELLO, U.S. Air Force (retired), *Co-Chair*
REBECCA NUGENT, Carnegie Mellon University, *Co-Chair*
PHILIP S. ANTON, Stevens Institute of Technology
TRILCE ESTRADA, University of New Mexico
MILLARD FIREBAUGH, University of Maryland (until December 2019)
STEPHEN FORREST, NAE,[1] University of Michigan
CHRISTINE FOX, Johns Hopkins University Applied Physics Laboratory
MELVIN GREER, Intel Corporation
CHARLES ISBELL, Georgia Institute of Technology
PETER LEVINE, Institute for Defense Analyses
ANN McKENNA, Arizona State University
ALYSON WILSON, North Carolina State University
JUN ZHUANG, University at Buffalo

Staff

LIDA BENINSON, Senior Program Officer, *Study Co-Director*
TYLER KLOEFKORN, Senior Program Officer, *Study Co-Director*
CARL-GUSTAV ANDERSON, Program Officer
STEVEN DARBES, Program Officer
SELAM ARAIA, Senior Program Assistant
ADRIANNA HARGROVE, Finance Business Partner (until December 2020)
HEATHER LOZOWSKI, Senior Finance Business Partner
BEN WATZAK, Research Assistant

MICHELLE K. SCHWALBE, Director, Board on Mathematical Sciences and Analytics

---

[1] Member, National Academy of Engineering.

BOARD ON MATHEMATICAL SCIENCES AND ANALYTICS
Division on Engineering and Physical Sciences

MARK L. GREEN, University of California, Los Angeles, *Chair*
HÉLÈNE BARCELO, Mathematical Sciences Research Institute
RUSSEL E. CAFLISCH, NAS,[1] New York University
DAVID S.C. CHU, Institute for Defense Analyses
RONALD R. COIFMAN, NAS, Yale University
JAMES (JIM) CURRY, University of Colorado, Boulder
RONALD D. FRICKER, JR., Virginia Polytechnic Institute and State University
SHAWNDRA HILL, Facebook
TRACHETTE JACKSON, University of Michigan
LYDIA KAVRAKI, NAM,[2] Rice University
TAMARA KOLDA, NAE,[3] Sandia National Laboratories
PETER KOUMOUTSAKOS, ETH Zurich
RACHEL KUSKE, Georgia Institute of Technology
JILL PIPHER, Brown University
YORAM SINGER, WorldQuant
TATIANA TORO, University of Washington
LANCE WALLER, Emory University
KAREN E. WILLCOX, University of Texas, Austin

Staff

MICHELLE K. SCHWALBE, Director
CARL-GUSTAV ANDERSON, Program Officer
SELAM ARAIA, Senior Program Assistant
LINDA CASOLA, Associate Program Officer (until December 2019)
ADRIANNA HARGROVE, Finance Business Partner (until December 2020)
TYLER KLOEFKORN, Senior Program Officer
HEATHER LOZOWSKI, Senior Finance Business Partner
BEN WATZAK, Research Assistant (until September 2020)

---

[1] Member, National Academy of Sciences.
[2] Member, National Academy of Medicine.
[3] Member, National Academy of Engineering.

COMMITTEE ON APPLIED AND THEORETICAL STATISTICS
Division on Engineering and Physical Sciences

NICHOLAS J. HORTON, Amherst College, *Co-Chair*
LANCE WALLER, Emory University, *Co-Chair*
F. DUBOIS BOWMAN, University of Michigan
ALICIA CARRIQUIRY, NAM,[1] Iowa State University
RONG CHEN, Rutgers University, The State University of New Jersey
WEI CHEN, NAE,[2] Northwestern University
MICHAEL J. DANIELS, University of Florida
AMY H. HERRING, Duke University
TIM HESTERBERG, Google, Inc.
REBECCA A. HUBBARD, University of Pennsylvania
KRISTIN LAUTER, Microsoft Research
DAVID MADIGAN, Columbia University
XIAO-LI MENG, Harvard University
RAQUEL PRADO, University of California, Santa Cruz
NANCY M. REID, NAS,[3] University of Toronto
CYNTHIA RUDIN, Duke University
AARTI SINGH, Carnegie Mellon University
ELIZABETH A. STUART, Johns Hopkins University
JONATHAN WAKEFIELD, University of Washington
ALYSON G. WILSON, North Carolina State University

Staff

TYLER KLOEFKORN, Director
CARL-GUSTAV ANDERSON, Program Officer
SELAM ARAIA, Senior Program Assistant
LINDA CASOLA, Associate Program Officer (until December 2019)
ADRIANNA HARGROVE, Finance Business Partner (until December 2020)
HEATHER LOZOWSKI, Senior Finance Business Partner
BEN WATZAK, Research Assistant (until September 2020)

MICHELLE K. SCHWALBE, Director, Board on Mathematical Sciences and Analytics

---

[1] Member, National Academy of Medicine.
[2] Member, National Academy of Engineering.
[3] Member, National Academy of Sciences.

AIR FORCE STUDIES BOARD
Division on Engineering and Physical Sciences

ELLEN M. PAWLIKOWSKI, NAE,[1] U.S. Air Force (retired), *Chair*
KEVIN G. BOWCUTT, NAE, The Boeing Company
CLAUDE R. CANIZARES, NAS,[2] Massachusetts Institute of Technology
MARK COSTELLO, Georgia Institute of Technology
WESLEY HARRIS, NAE, Massachusetts Institute of Technology
JAMES E. HUBBARD JR., NAE, Texas A&M University
LESTER L. LYLES, U.S. Air Force (retired)
WENDY M. MASIELLO, U.S. Air Force (retired)
LESLIE ANN MOMODA, HRL Laboratories, LLC
OZDEN OCHOA, Texas A&M University
HON F. WHITTEN PETERS, Williams & Connolly, LLP
HENDRICK RUCK, Edaptive Computing, Inc.
JULIE J.C.H. RYAN, Wyndrose Technical Group
MICHAEL D. SCHNEIDER, Lawrence Livermore National Laboratory
GRANT H. STOKES, MIT Lincoln Laboratory

Staff

ELLEN CHOU, Director
GEORGE COYLE, Senior Program Officer
RYAN MURPHY, Program Officer
ADRIANNA HARGROVE, Finance Business Partner
MARGUERITE SCHNEIDER, Administrative Coordinator
EVAN ELWELL, Research Assistant

---

[1] Member, National Academy of Engineering.
[2] Member, National Academy of Sciences.

COMPUTER SCIENCE AND TELECOMMUNICATIONS BOARD
Division on Engineering and Physical Sciences

LAURA HAAS, NAE,[1] University of Massachusetts, Amherst, *Chair*
DAVID E. CULLER, NAE, University of California, Berkeley
ERIC HORVITZ, NAE, Microsoft Research
CHARLES ISBELL, Georgia Institute of Technology
ELIZABETH MYNATT, Georgia Institute of Technology
CRAIG PARTRIDGE, Colorado State University
DANIELA RUS, NAE, Massachusetts Institute of Technology
FRED B. SCHNEIDER, NAE, Cornell University
MARGO I. SELTZER, NAE, University of British Columbia
NAMBIRAJAN SESHADRI, NAE, University of California, San Diego
MOSHE Y. VARDI, NAS[2]/NAE, Rice University

Staff

JON EISENBERG, Senior Board Director
SHENAE BRADLEY, Administrative Assistant
RENEE HAWKINS, Financial and Administrative Manager
LYNETTE I. MILLETT, Associate Director
KATIRIA ORTIZ, Associate Program Officer
BRENDAN ROACH, Program Officer

---

[1] Member, National Academy of Engineering.
[2] Member, National Academy of Sciences.

BOARD ON HIGHER EDUCATION AND WORKFORCE
Policy and Global Affairs

KUMBLE R. SUBBASWAMY, University of Massachusetts, Amherst, *Chair*
ANGELA BYARS-WINSTON, University of Wisconsin, Madison
JAIME CURTIS-FISK, The Dow Chemical Company
MARIELENA DESANCTIS, Broward College
APRILLE J. ERICSSON, NASA Goddard Space Flight Center
JOAN FERRINI-MUNDY, University of Maine
GABRIELA GONZALEZ, NAS,[1] Louisiana State University
TASHA R. INNISS, Spelman College
SALLY K. MASON, University of Iowa
DOUGLAS S. MASSEY, NAS, Princeton University
RICHARD K. MILLER, NAE,[2] Olin College of Engineering
PATRICIA SILVEYRA, University of North Carolina at Chapel Hill
KATE STOLL, Massachusetts Institute of Technology Washington Office
MEGHNA TARE, University of Texas
MARY WOOLLEY, NAM,[3] President and CEO, Research! America

Staff

LEIGH MILES JACKSON, Acting Director
THOMAS RUDIN, Director (until November 2020)
AUSTEN APPLEGATE, Senior Program Assistant
ARIELLE BAKER, Associate Program Officer
ASHLEY BEAR, Senior Program Officer
LIDA BENINSON, Senior Program Officer
FRAZIER BENYA, Senior Program Officer
IMANI BRAXTON-ALLEN, Senior Program Assistant (until April 2021)
MARIA LUND DAHLBERG, Senior Program Officer
ALEX HELMAN, Program Officer
REBEKAH HUTTON, Program Officer
PRIYANKA NALAMADA, Associate Program Officer
LAYNE A. SCHERER, Senior Program Officer
JOHN VERAS, Senior Program Assistant
MARQUITA WHITING, Senior Program Assistant

---

[1] Member, National Academy of Sciences.
[2] Member, National Academy of Engineering.
[3] Member, National Academy of Medicine.

COMMITTEE ON NATIONAL STATISTICS
Division of Behavioral and Social Sciences and Education

ROBERT M. GROVES, NAS[1]/NAM,[2] Georgetown University, *Chair*
LAWRENCE D. BOBO, NAS, Harvard University
ANNE C. CASE, NAS/NAM, Princeton University
MICK P. COUPER, University of Michigan
JANET CURRIE, NAS/NAM, Princeton University
DIANA FARRELL, JPMorgan Chase Institute, Washington, DC
ROBERT GOERGE, University of Chicago
ERICA L. GROSHEN, Cornell University
HILARY HOYNES, University of California, Berkeley
DANIEL KIFER, Pennsylvania State University
SHARON LOHR, Arizona State University
JEROME P. REITER, Duke University
JUDITH A. SELTZER, University of California, Los Angeles
C. MATTHEW SNIPP, Stanford University
ELIZABETH A. STEWART, Johns Hopkins Bloomberg School of Public Health
JEANETTE M. WING, Columbia University

Staff

BRIAN HARRIS-KOJETIN, Director
TARA BECKER, Program Officer
CONSTANCE F. CITRO, Senior Scholar
MICHAEL L. COHEN, Senior Program Officer
DANIEL L. CORK, Senior Program Officer
STUART ELLIOTT, Scholar
MARY GHITELMAN, Senior Program Assistant
NANCY J. KIRKENDALL, Senior Program Officer
REBECCA KRONE, Program Coordinator
CHRISTOPHER D. MACKIE, Senior Program Officer
MALAY MAJMUNDAR, Senior Program Officer
ANTHONY S. MANN, Program Associate
KRISZTINA MARTON, Senior Program Officer
MICHAEL J. SIRI, Associate Program Officer
JORDYN WHITE, Program Officer

---

[1] Member, National Academy of Sciences.
[2] Member, National Academy of Medicine.

Preface

The ability to use analytical tools to collect and analyze data is rapidly expanding and bringing along new challenges and opportunities. Following the National Academies of Sciences, Engineering, and Medicine report on *Data Science for Undergraduates: Opportunities and Options* and Roundtable on Data Science Postsecondary Education—and in recognition of the rapidly evolving premium on data and its power—the Department of Defense (DoD) requested this study to determine how it could prepare its acquisition workforce to embrace and exploit the rapidly evolving promise of data science.

To carry out the task, the National Academies appointed a committee of 13 members. The committee was extraordinary in its makeup, with experts in defense acquisition processes, policies, analyses, and workforce management; leading industry data scientists; and academic experts in engineering, mathematics, statistics, computer science, data science, and education.

Over the course of the study, the committee came to appreciate the complexity and diversity of the DoD acquisition and personnel systems and processes, as well as the multitude of ways industry and government counterparts have addressed the upskilling of their workforces. In the end, the committee concluded that embracing data science and the possibilities for insight at the acquisition level must be driven by leadership. It will take time to upskill the entire acquisition workforce, but the tools, training, and supplemental skills are available today. Deliberate commitment, resources, and clear priorities across DoD will be required to make it happen.

We thank all of the members of the committee for their intensive effort and collaborative spirit in crafting this report. We were aided by remarkably talented co-study directors, Lida Beninson and Tyler Kloefkorn, and an able group of staff, including Carl-Gustav Anderson, Selam Araia, Steven Darbes, Adrianna Hargrove, Heather Lozokowski, Katiria Ortiz, Michelle Schwalbe, and Ben Watzak. It was our pleasure and honor to be a part of this study.

We look forward to seeing the DoD acquisition community harness data to address today's challenges and opportunities as well those on the horizon, both in the acquisition process and in the products that system delivers.

> Wendy M. Masiello and Rebecca Nugent, *Co-Chairs*
> Committee on Improving Defense Acquisition
> Workforce Capability in Data Use

Acknowledgments

The committee would like to thank the following individuals for providing input to this study.

Brij Agrawal, Naval Postgraduate School
Darryl Ahner, Air Force Institute of Technology
Jonathan Alt, Naval Postgraduate School
Darin Ashley, Air Force Contracting Center
Mike Baylor, Lockheed Martin
Bethany Blakey, General Services Administration
Gary Bliss, Institute for Defense Analyses
Dan Boger, Naval Postgraduate School
Stacy Bostjanick, Office of the Under Secretary of Defense for Acquisition and Sustainment
John Bottega, EDMCouncil
Steve Brady, Air Force Contracting Center
Mark Breckenridge, Department of Defense Office of People Analytics
Yisroel Brumer, Cost Assessment and Program Evaluation
David Cadman, Department of Defense
Matthew Carlyle, Naval Postgraduate School
Yu-Chu Chen, Naval Postgraduate School
Michael Conlin, Department of Defense
Stacy Cummings, Department of Defense
Jesse Cunha, Naval Postgraduate School
Robert Dell, Naval Postgraduate School
Lisa Disbrow, Johns Hopkins University Applied Physics Laboratory

Bess Dopkeen, U.S. House Committee on Armed Services
Eric Eckstrand, Naval Postgraduate School
Tammy Foster, Lockheed Martin
Michael Gilmore, Institute for Defense Analyses
Vance Gilstrap, Defense Acquisition University
James Greene, Naval Postgraduate School
Mary Hauber, Air Force Contracting Center
Pete Herrmann, Air Force Contracting Center
Hans Jerrell, Defense Acquisition University
Sallie Ann Keller, University of Virginia
Mark Krzysko, Department of Defense
Bruce Litchfield, Lockheed Martin
Megan McKernan, RAND Corporation
Sears Merritt, MassMutual
Robert Mortlock, Naval Postgraduate School
William Muir, Naval Postgraduate School
Paul Nielsen, Optum Technologies
Walter Owen, Naval Postgraduate School
Sezin Palmer, Johns Hopkins University Applied Physics Laboratory
William Parker, Defense Acquisition University
Renee Pasman, Lockheed Martin
Matthew Rattigan, University of Massachusetts Amherst
Rene Rendon, Naval Postgraduate School
David Robinson, Defense Acquisition University
Thomas Sasala, U.S. Navy
Clyde Scandrett, Naval Postgraduate School
James Scrofani, Naval Postgraduate School
Garry Shafovaloff, Human Capitals Initiative
Heidi Shyu, VK Integrated Systems
Jaeki Song, Texas Tech University
Nancy Spruill, Department of Defense (ret.)
Luis Stevens, Target Corporation
Carol Tisone, Defense Acquisition University
Darlene Urquhart, Defense Acquisition University
Eileen Vidrine, U.S. Air Force
Maryann Watson, Defense Acquisition University
Rebecca Weirick, Office of the Deputy Assistant Secretary of the Army
Roger Westermeyer, Air Force Contracting Center

Acknowledgment of Reviewers

This Consensus Study Report was reviewed in draft form by individuals chosen for their diverse perspectives and technical expertise. The purpose of this independent review is to provide candid and critical comments that will assist the National Academies of Sciences, Engineering, and Medicine in making each published report as sound as possible and to ensure that it meets the institutional standards for quality, objectivity, evidence, and responsiveness to the study charge. The review comments and draft manuscript remain confidential to protect the integrity of the deliberative process.

We thank the following individuals for their review of this report:

Ben Baumer, Smith College,
Wesley Bush, NAE, Northrop Grumman Corporation,
David Chu, Institute for Defense Analyses,
Bakari Dale, Office of the Secretary of the Army-OBT,
Robert Dell, Naval Postgraduate School,
Ronald Fricker, Virginia Tech,
Sallie Ann Keller, NAE, University of Virginia,
General Lester Lyles, NAE, Independent Consultant,
Brandeis Marshall, Spelman College,
Katharina McFarland, Independent Consultant,
Cherry Murray, NAS/NAE, Harvard University (emerita) and University of Arizona,
Fred Oswald, Rice University,
The Honorable Sean J. Stackley, L3Harris Technologies,

Kristin Tolle, University of Washington, and
Lawrence Woolf, General Atomics Aeronautical Systems.

Although the reviewers listed above provided many constructive comments and suggestions, they were not asked to endorse the conclusions or recommendations of this report nor did they see the final draft before its release. The review of this report was overseen by Stephen M. Robinson, NAE, University of Wisconsin, Madison. He was responsible for making certain that an independent examination of this report was carried out in accordance with the standards of the National Academies and that all review comments were carefully considered. Responsibility for the final content rests entirely with the authoring committee and the National Academies.

Contents

SUMMARY		1
1	**INTRODUCTION**	6
	Harnessing Data for Defense Acquisition, 7	
	Leadership Driving Change in a Data-Centric Organization, 12	
	Study Origin, Objectives, and Approach, 15	
	Report Outline, 16	
	References, 17	
2	**DEFENSE ACQUISITION PROCESS, DATA, AND WORKFORCE: THE SHORT VERSION**	19
	The Acquisition Process, 19	
	Acquisition Data Use and Context, 21	
	The Acquisition Workforce, 22	
	References, 23	
3	**DATA SCIENCE AND THE DATA LIFE CYCLE: THE SHORT VERSION**	24
	What Is Data Science and Who Does It?, 24	
	Data Science Is Collaborative and Cyclical, 24	
	The Data Life Cycle and Its Phases, 28	
	Data Ethics, Privacy, and Security, 31	
	References, 32	

4	DATA SCIENCE IN DOD ACQUISITION	33

Opportunities for Improved Data Use in Defense Acquisition, 33
Defense Acquisition Functions and the Data Life Cycle, 35
References, 41

5	DATA LIFE CYCLE MINDSET, SKILLSET, TOOLSET: ROLES AND TEAMS	42

Mindset, 42
Skillset, 46
Toolset, 52
Team Structures for Data Science, 54
References, 60

6	PREPARING AND SUSTAINING A DATA-CAPABLE DEFENSE ACQUISITION WORKFORCE	62

Overview of Training Approaches in Data Science and Analytics, 63
Data Use and Analysis Training Opportunities for Defense Acquisition Personnel—Current Programs, 68
Industry and Government Training Approaches in Data Science and Analytics, 73
Envisioned Future for Training in Data Use Capabilities for Defense Acquisition Personnel, 75
References, 76

7	FINDINGS, CONCLUSIONS, AND RECOMMENDATIONS	78

Data Science in DoD Acquisition, 78
Data Life Cycle Mindset, Skillset, Toolset: Roles and Teams, 79
Preparing and Sustaining a Data-Capable Workforce, 81
Synopsis, 82

APPENDIXES

A	Meeting and Workshop Agendas	85
B	Defense Acquisition Notes	95
C	Data Science Case Studies in Defense Acquisition	109
D	Skills for Data Science Mastery	120
E	Glossary of Terms, Abbreviations, and Acronyms	124
F	Committee Member Biographies	127

Summary

Data are a strategic asset. The effective use of data science—the science and technology of extracting value from data—improves, enhances, and strengthens acquisition decision-making and outcomes. Those outcomes include improvements in missions, portfolios of capabilities, and the schedules, costs, and performance of acquired systems for net improvements to warfighters, operators, and supporting organizations. Using data science to support decision making is not new to the defense acquisition community; its use by the acquisition workforce has enabled acquisition and thus defense successes for decades. Still, more consistent and expanded application of data science will continue improving acquisition outcomes, and doing so requires coordinated efforts across the defense acquisition system and its related communities and stakeholders. Central to that effort is the development, growth, and sustainment of data science capabilities across the acquisition workforce.

The Under Secretary of Defense for Acquisition and Sustainment [USD(A&S)] tasked the Committee on Improving Defense Acquisition Workforce Capability in Data Use to assess how data science can improve acquisition processes and develop a framework for training and educating the defense acquisition workforce to better exploit the application of data science. Key to the framework's development was the identification of opportunities where data science can improve acquisition processes, the relevant data science skills and capabilities necessary for the acquisition workforce, and relevant models of data science training and education.

CONCLUSIONS AND RECOMMENDATIONS

Continue Improving Defense Acquisition with Data Science

There are examples of the successful application of data science in Department of Defense (DoD) acquisition. There are also clear opportunities to better use data science across acquisition policy making, processes, decision making, and actions. Data science has improved acquisitions processes, for example, by enhancing program and contracting analysis, tracking cost and program performance, enabling analysis of alternatives, determining whether a weapon or platform meets performance requirements, and informing decision making. Typically, improvements can be made in the collecting, curating, and managing phases of the data life cycle for broad range of acquisition functions, including cost estimating, contracting, contract management and cybersecurity. Currently, there are several shortcomings in the front end of data science processes in defense acquisition—namely in data collection and curation. Also, data silos can hinder shared data across DoD and are a common obstacle preventing the utilization of the full potential of the data life cycle. The vision and opportunity for data science in the defense acquisition system is one in which data collection and use is not a support function but is integrated into all acquisition processes.

The Secretary of Defense and other senior leaders can facilitate adoption by *demanding* high-quality, complete, and accurate data for their decisions and those below them while providing the necessary financial and personnel resources for obtaining these data and analyses. This pull is necessary to drive behavior and complement the push from technical opportunities, putting action behind their strategic statements.

> **Recommendation 1.1:** The Department of Defense (DoD) and Congressional leaders and stakeholders should promote the cultural changes necessary to facilitate data-informed decisions across the entire DoD by insisting that high-quality, complete, and accurate data and analysis be provided to inform their own decisions and those of their subordinates.

> **Recommendation 4.1:** The Department of Defense should continue to seek improvements in defense acquisition through the increased application of data science, including addressing shortfalls in data collection, curation, management, and sharing.

Opportunities Exist to Improve Workforce Abilities Across the Data Science Life Cycle

Extracting value from data requires a collective data science mindset, skillset, and toolset. Central to data science is the data life cycle, which is an iterative, bi-directional workflow—with people at each element—that has several phases: questions, definitions, coordination, generation of data, collection of data, curation and management of data, analysis of data, dissemination and interpretation, and assessment.

The committee found that an increasing majority of the acquisition workforce is participating in the data life cycle but may not recognize their roles and value within it. Understanding their roles across the data life cycle is critical for improved and sustained data use and data-informed decision-making. For utilizing the full potential of the data life cycle, teams will need to be established, customized, nurtured, and managed. Team leaders will also need familiarity with the data life cycle, management skills that support and optimize data science talent, and a commitment to data-informed decision-making. Managing data science projects uses strategies and approaches for leading collaborative, cross-functional, technical projects. Specific attention is necessary for the development of a team with skills across the data life cycle and the ability to ask questions specific to the quality and utility of data.

> **Recommendation 5.2:** The Department of Defense should prioritize the utilization of data and the data life cycle by appropriate and judicious investment in the acquisition workforce data science mindset, skillset, and toolset.

Data Literacy Skills Are Important Throughout the Acquisition Workforce

DoD and its components should ensure that all members of the acquisition workforce have at least basic (non-technical) data science skills, which include an understanding of the data life cycle and how it works, an ability to consume and understand data story-telling and data-informed decisions, and an ability to recognize matters of ethics, privacy, and security. These skills make up what is often called *data literacy*—a set of data-centric skills that have evolved with data science and are growing in importance and use in government, industry, and academia.

> **Recommendation 5.1:** The Department of Defense and its components should ensure that all members of the acquisition workforce and its leadership eventually have data literacy—the basic, non-technical data science skills that include an understanding of the data life cycle and

how it works, an ability to consume and understand data story-telling and data-informed decisions, and an ability to recognize matters of ethics, privacy, and security—all skills that evolve along with the field and its use in government, industry, and academia.

Advanced Capabilities Can Enable Improved Data Use in Acquisition

Collectively, the defense acquisition workforce must have capabilities in all phases of the data life cycle—from data collection and curation to data analysis and visualization—to fully exploit the value of a particular data stream and its corresponding domain application. Executing the data life cycle is a collaborative endeavor, requiring a collective skillset found in teams of data engineers, data scientists, data analysts, data users, domain experts, and leaders/decision makers. Data scientists are experts across the data life cycle, with special emphasis on advanced techniques for curation, management, analysis, visualization, dissemination, and interpretation. The vast majority of the defense acquisition workforce will identify as domain experts, leaders/decision makers, and even data analysts and data users (under the guise of their acquisition job functions), but data engineers and data scientists are also critical for improved data use.

> **Finding 5.6:** Executing the data life cycle is a collaborative endeavor and generally requires a collective skillset found in teams of data engineers, data scientists, data analysts, data users, domain experts, and leaders/decision makers.

Diverse, Tailored, and Situated Training Models Can Increase Data Capabilities and Outcomes in the Acquisition

Workforce development in data science is critical to the success of DoD's current and future acquisition improvement efforts. The committee examined a variety of training efforts within and outside of DoD for increasing data capabilities of the defense acquisition workforce. Currently, there are several non-DoD data science training efforts that could be leveraged for the acquisition workforce; some are already being used, but use could be expanded. To be effective, data science training needs to be applied in the context of acquisition functions—preferably with realistic data and examples—rather than simply taught in the abstract without applications. Also, there is no one-size-fits-all training approach; training must be tailored to meet specific needs depending on professional roles, the application and scale of data, and mission goals and priorities. In addition, as data science curricula and courses are piloted for the acquisition community, clear metrics are key for evaluating the success and applicability of the training approaches. Successful training efforts can be scaled to increase impact and broaden the data capabilities of the defense acquisition workforce.

Recommendation 6.1: Institutions that provide training for defense acquisition professionals should ensure that courses integrate realistic data and challenges in currently available and future courses, including non-data-focused courses. Realistic data and challenges offer students the opportunity to learn about uncertainty, sampling, variability, and noise. This realism provides personnel the opportunities to learn and apply data techniques in acquisition scenarios and projects.

Recommendation 6.2: The defense acquisition system should continue to leverage and expand the variety of data science training options available both within and outside of the Department of Defense for the acquisition workforce. Trends include higher education, certificate programs, data "bootcamps," micro-credentialing, online courses, and executive and leadership training. These options include data analytics and data science training offered by the Military Services, internal and external higher education institutions, and the commercial sector.

Recommendation 6.3: New training programs for acquisition leadership and personnel in data science should incorporate assessment metrics to evaluate their success. Successful efforts should then be expanded to increase effectiveness and broaden the data capabilities of the defense acquisition workforce.

Recommendation 6.4: Defense acquisition leadership should take training goals and characteristics into account when selecting or developing data science training. Clarifying the data goals and key data capabilities necessary for different acquisition teams are essential steps for identifying suitable training programs.

Recommendation 6.5: Collectively, these recommendations should be considered by the defense acquisition leadership as part of the larger Department of Defense (DoD) Data Strategy. While DoD has a data strategy, it lacks specific guidance for defense acquisition. Specific guidance should address not only acquisition data governance (e.g., collection, curation, management, and access), analysis, and consumption but also the workforce that facilitates these functions and is central to the data life cycle. Tradeoffs and investment limitations abound, so a strategic plan is critical to guiding and ensuring prioritized investments to maximize payoff.

1

Introduction

In September of 2020, the Department of Defense (DoD) published a data strategy (DoD 2020a), intended to support the National Defense Strategy and Digital Modernization program. While not focusing on or mentioning acquisition explicitly, the DoD Data Strategy (summarized in Figure 1.1) includes guiding principles, data goals, and essential capabilities needed to meet those goals. The strategy envisions a DoD that is a "data-centric organization that uses data at speed and scale for operational advantage and increased efficiency." (p. 2). The strategy frames data as a strategic asset and concludes that "every leader must treat data as a weapon system, stewarding data throughout its lifecycle and ensuring it is made available to others" (p. 11). See Box 1.1. However, successful implementation of a data strategy requires the support of leadership across all aspects of the organization, adequate funding, and a trained workforce. While the DoD Data Strategy is in the early phase of deployment, the importance of data as a strategic resource and enterprise asset for DoD cannot be understated.

In 2019, the Office of Under Secretary of Defense for Acquisition and Sustainment (USD(A&S)) recognized the rapidly expanding tools and techniques available in data science and commissioned this report to identify how DoD can enable the defense acquisition system and its workforce to make better use of data science. To that end, this report explores the critical roles for data within the defense acquisition system, the importance of aligning and operationalizing a data strategy to improve acquisition processes and insights, and approaches toward developing an acquisition workforce better able to apply data-driven methods to their work and interpret their results. Data alone will not remedy many of the complex problems facing

INTRODUCTION

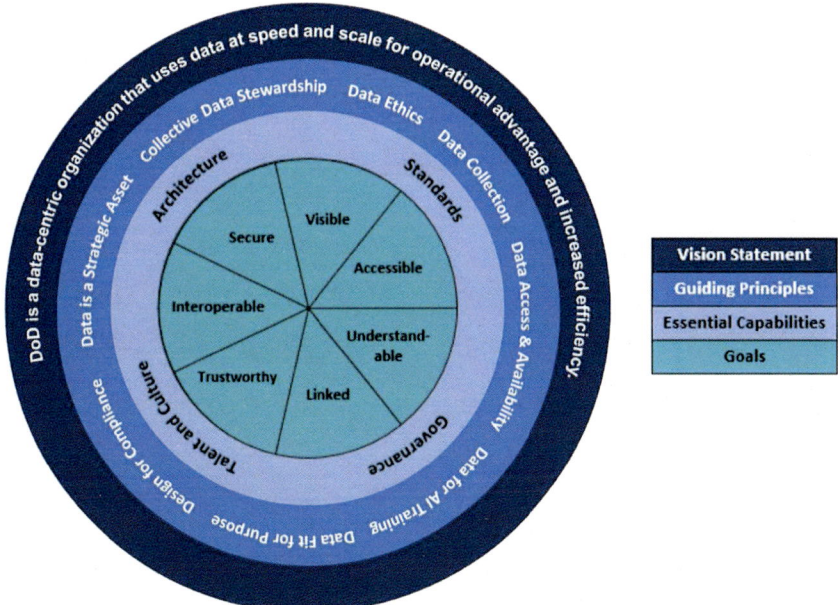

FIGURE 1.1 DoD Data Strategy. SOURCE: Department of Defense. "DoD Data Strategy." Department of Defense. 2020. https://media.defense.gov/2020/Oct/08/2002514180/-1/-1/0/DOD-DATA-STRATEGY.PDF.

the acquisition system, nor does every acquisition professional need to become a data scientist or data engineer. Nevertheless, improved appreciation, sharing, use, and analysis of data along with data-informed decision making have the potential to dramatically improve overall efficiency and effectiveness of defense acquisition. This report conveys existing opportunities to leverage advances in data science for defense acquisition. As data science and its tools are rapidly evolving, DoD will need to continually refine and update its understanding of data science opportunities and tailor them to defense acquisition.

HARNESSING DATA FOR DEFENSE ACQUISITION

Data science—the multidisciplinary field that encompasses the technologies, processes, and systems to extract knowledge and insight from data and enable data-informed decision making under uncertainty, to be discussed later in this report—has become a pervasive and transformative field unto itself over the past decade. Building a skilled workforce that is able to apply and adapt to the quickly evolving domain of data science

BOX 1.1
DoD Data Strategy Goals

The *DoD Data Strategy* states that data should be visible, accessible, understandable, linked, trustworthy, interoperable, and secure (abbreviated as VAULTIS). Each of these seven broad goals is relevant to the defense acquisition system.

- *Visible* data "enables authorized users to discover the existence of data that is of particular interest or value. Data stewards, data custodians, and functional data managers are all responsible and obligated to make their data visible to authorized users by identifying, registering, and exposing data in a way that makes it easily discoverable across the enterprise, and to external partners as appropriate."
- *Accessible* data "enables authorized users to obtain the data they need when they need it, including having data automatically pushed to interested and authorized users...DoD is making data, including warfighting, intelligence, and business data, accessible to authorized users. Accessibility requires that protective mechanisms (e.g., security controls) are in place for credentialed users to ensure that access is permitted in accordance with laws, regulations, and policies."
- *Understandable* data "is critical to enable enhanced, more accurate, and timely decision-making. The inability to aggregate, compare, and truly understand data adversely affects the ability of the Department to react and respond. Without proper context, interpretation and analysis of the data could be flawed, resulting in potentially fatal outcomes. Bringing together business and technology and applying a data-centric approach enable massive amounts of data to be transformed into the insights needed to lead DoD more effectively and efficiently."
- *Linked* data ensures "that relationships and dependencies can be uncovered and maintained. Adhering to industry best-practices for open data standards, data catalogs, and metadata tagging, the Department ensures that connections across disparate sources can be made and leveraged for analytics."
- *Trustworthy* data enables the delivery of "needed value to its Service members, civilians, and stakeholders. Lacking confidence in the data may result in less timely decision-making or, consequently, no decision when one is warranted."
- *Interoperable* data are "critical for successful decision-making and joint military operations. Achieving semantic as well as syntactic interoperability using common data formats and machine-to-machine communications accelerates advanced algorithm development and provides a strategic advantage to the Department."
- *Secure* data "allows DoD to maximize the use of data while, at the same time, employing the most stringent security standards to protect the American people."

SOURCE: Department of Defense, 2020, *DoD Data Strategy*, September 30, https://media.defense.gov/2020/Oct/08/2002514180/-1/-1/0/DOD-DATA-STRATEGY.PDF.

is essential for organizations looking to improve performance through more optimal use of data. A skilled workforce—one capable of extracting value from data—will include a number of different data science roles and teamwork among them. Some team members will have technical expertise in data science, while others are more specific to the "domain" (contracting, acquisition, materials, testing, cyber, etc.). All are more fully described later in this report.

With respect to data scientists specifically, private industry holds a significant advantage in recruiting and retention, in part because of the high salaries offered and the flexibility of these organizations to more rapidly adopt new approaches. In comparison, currently DoD and other government organizations have a more limited toolset available to recruit, train, and retain skilled data scientists. As DoD acquisition contemplates the use of data science teams in support of its acquisition workforce, the cost and challenges of recruiting sophisticated data scientists may be a driver in how many and where such teams are established and the mix of talent between military, civil service, and contracted specialists.

In spite of the recruiting and retention challenges in the data science, an increasing number of the defense acquisition teams are embracing aspects of data science—in particular, the use of data and data analysis—and are reaping the benefits. The USD(A&S) has an important mission of developing policy and processes to enable more effective and efficient acquisition by the DoD Components[1] that result in operational benefits to warfighters. With the potential to provide products for rapidly evolving missions, accelerate logistics, lower costs, and save lives, both the Office of the Secretary of Defense (OSD) and the Components are striving to collect, manage, and make better use of data. Examples are described in Box 1.2.

The successes of teams—such as those cited in Box 1.2—are encouraging, yet even these successes had their share of challenges with access to data, modern tools, and talent. Studies of existing DoD acquisition data systems often cite the enormity, complexity, and siloed nature of data within the Defense community and the related challenges these issues cause. For example, in the January 2019 Report of the Advisory Panel on Streamlining and Codifying Acquisition Regulations (Section 809 panel), Recommendation 89 identified the "[t]he proliferation of different data architectures throughout the acquisition and financial system leads to countless marginal inefficiencies in DoD," and recommending that Congress "Direct DoD to

---

[1] DoD Components are the Office of the Secretary of Defense, Office of the Chairman of the Joint Chiefs of Staff and the Joint Staff, Combatant Commands, Office of Inspector General, Military Departments, Defense Agencies, DoD field activities, and other organizational entities, which includes the National Guard Bureau (DoD 2020b).

> **BOX 1.2**
> **The Power of Data: Examples in Defense Acquisition**
>
> **Enabling Multi-Year Appropriation Approval and Program Benchmarking**
>
> The Cost Assessment Data Enterprise (CADE) system (operated by the Office of the Secretary of Defense's (OSD's) Cost Assessment and Program Evaluation (CAPE) directorate) is a digitized database and analysis toolset that provides high-quality, accessible, and structured historical cost data from contractors who developed previous DoD weapon systems. Through CADE, cost analysts in CAPE and throughout DoD are able to perform program cost estimation using validated and consistent data. These analytic results provide rigorous cost benchmarks for acquisition programs, enabling costs comparisons, affordability analysis, and more reliable cost estimates and budgeting. By having reliable benchmarks, DoD program managers and planners are in a stronger position to request realistic budgetary profiles and negotiate prices with prime contractors, sometimes resulting in billions of dollars of savings and cost avoidance.
>
> **Saving Billions of Dollars with Category Management**
>
> In 2014, the Air Force published its Category Management Concept of Operations (CONOPS), which is an industry best practice that helps organizations (such as Walmart, UPS, and Kroger) segment their spending into areas that contain similar or related products. The Category Management CONOPS enables clarity on total dollars spent in any category, a market assessment on how the commercial market actually buys and manages each category, and strategic Air Force-level decisions on how best to buy a category of products or services. Results of a category study might be a new approach of buying that is more consistent with the private industry, a decision to centrally or de-centrally buy the category, or a decision to stop buying something entirely and redistribute needed funds to higher priority needs. Key to this success is the availability of procurement data to enable the creation of relevant categories and the integrated management to improve outcomes. Supported by a centralized data science team under the Air Force Installation Contracting Center, the Category Management CONOPS has recorded more than $2 billion in savings for the Air Force between 2016 and mid-2019.[a,b]
>
> **Improving Logistics and Operational Availability of Military Equipment**
>
> The U.S. Army's Velocity Management initiative monitors the time, cost, and resulting quality for fulfilling parts and items orders for maintenance. These per-

consolidate or eliminate competing data architectures within the defense acquisition and financial system" (DoD 2019, p. 1).

As a result of the proliferation of data architectures and long-standing challenges in data labeling and terminologies, the DoD acquisition system often misses opportunities to share and contextualize data, account for un-

formance metrics and their associated data helped to inform the availability and maintenance status of military equipment. Together with other key elements (e.g., leadership commitment, higher fulfilment from the main depot, and direct deliveries to customers), this initiative has expedited parts delivery, shortened maintenance time, and increased the operational availability of military equipment for warfighters, putting military equipment back into the field in significantly less time. Data-driven measurement, metric feedback, and new information systems—along with simplified performance rules—were key elements of Velocity Management.[c]

Saving Lives with Test and Analysis Programs

Operational test and evaluation are data-driven processes that examine how new or upgraded systems will perform under realistic combat conditions prior to full-rate production and fielding to combat units. As one example, beginning in 2007, congressional concern about body armor testing led to Director Operational Test and Evaluation (DOT&E) working closely with the military services to develop test protocols for the ballistic components of combat helmets. Applying statistical design of experiments at a program level allowed DoD to establish a testing and analysis program that identified that an initial design for the Enhanced Combat Helmet did not provide sufficient protection against small-arms penetrations and led to a design change in the ballistic shell laminate material.[d,e] Key to this success is the collection and analysis of test data. This example illustrates not only the value of data science but also how its application is inherently embedded within an acquisition discipline.

---

[a] R. Westermeyer, 2017, "Bullet Background Paper on Air Force Installation Contracting Agency's Business Analytics Capability," Wright-Patterson Air Force Base, Ohio.
[b] R. Westermeyer, 2019, "Bullet Background Paper on Business Reform—Category Management." Wright-Patterson Air Force Base, Ohio.
[c] J. Dumond, and R. Eden, 2005, "Improving Government Processes: From Velocity Management to Presidential. Appointments," RAND Corporation, https://www.rand.org/pubs/reprints/RP1153.html.
[d] Director Operational Test and Evaluation, 2016, "Director, Operational Test and Evaluation (DOT&E) FY 2016 Annual Report, https://www.dote.osd.mil/Publications/Annual-Reports/2016-Annual-Report/.
[e] J. Hester, T. Johnson, and L. Freeman, 2015, "Managing Risks: Statistically Principled Approaches to Combat Helmet Testing," in *Research Notes,* Fall, Institute for Defense Analyses, Alexandria, VA.

certainties, and consistently incorporate data use and analysis into decision-making processes. Data use for acquisition insights has been siloed within disciplines and organizational elements, and widespread data sharing and use of more sophisticated tools has been limited. Moreover, the pool of data science experts that understand defense acquisition is small. These data-related challenges are common across government organizations and all sectors of industry. However, due to the size and complexity of the DoD acquisition system, described later in this report, it can be challenging to determine a universal approach for improving data use. In general, though, extracting decision quality information from data will depend on use of the data life cycle, to be described later in this report.

A comprehensive *data strategy*, specific to USD(A&S), could help move beyond pockets of innovation and meet the need for data-informed decision making throughout the defense acquisition system. Because the defense acquisition system relies on data generated and maintained throughout DoD, a successful implementation of a data strategy would guide which data to generate and collect and how that data should be stored, managed, shared, and used. As USD(A&S) determines its data strategy consistent with the 2020 DoD Data Strategy framework (Figure 1.1 and Box 1.2), it needs to frame, inform, prioritize, and guide all aspects of the data life cycle in order to fully capitalize on the value of data already generated within the acquisition system, as well as tap into and feed data into other DoD data systems.

LEADERSHIP DRIVING CHANGE IN A DATA-CENTRIC ORGANIZATION

Chief data officers (CDOs) for OSD and the military departments play a key role in the implementation of the DoD Data Strategy. Brought in prior to the release of the strategy, these CDOs are collectively seeking order and discipline in data use.

During the April 2020 National Academies workshop on "Improving Defense Acquisition Workforce Capability in Data Use," Mr. Tom Sasala, the chief data officer for the Department of the Navy, noted that two of the Navy's key challenges are that (1) data are not available for mission commanders, warfighters, and decision makers, and (2) data silos limit real-time use of data and linked data, and force duplication of efforts (Sasala 2020). As CDO of the Navy, he stated that much of his role is to make data available to address the challenges of accessibility, sharing, and security. "Our goal is just to maximize access to tools and data and allow people to be innovative." Current priorities for Navy include data specifications, common platform development, and data integration, which are the data engineering tasks that prepare data for use in analysis and decision making (Sasala 2020).

Ms. Eileen Vidrine, CDO of the Air Force, shared her office's goal to "make data part of every airman's core DNA" (Vidrine 2020). The Air Force CDO oversees a small laboratory based at Andrews Air Force Base. This lab considers topics proposed by Air Force personnel and develops teams composed of a data scientist, data architect, and a subject-matter expert to work on small use cases that could have enterprise-level impact. In their work, the lab identifies certain data sets that have been required multiple times and prioritizes their cleaning, sharing, and accessibility (Vidrine 2020). Additionally, the lab discovered useful data at the tactical level that are stored on an individual's desktop.

In a January 2020 interview, Vidrine explained, "In our first year, we just wrapped up our 15th use case, and several of them are now being scaled across the enterprise. We take a problem set, we use data science to solve it, and then we use that as the seed to grow the next use case moving forward" (Miller 2020). A key goal is to leverage cloud infrastructure to allow the airmen to register the data and make them discoverable so that the data can support broader strategic questions.

The DoD CDOs operate at the Service level, not specifically in the acquisition lane. But the data fundamentals are the same: good, clean, shareable data.

The long-term vision and opportunity for data science is integrating it into operations rather than just a side support function providing insight. As one moves from the concept of data science as a separate activity to one where the use and analysis of data are fully integrated within an organization's operations, management, and decision making, system interoperability not only facilitates data *sharing* (to enable analysis) but also *data-informed decision making and operations*. To achieve this, system interoperability, data sharing, and data accessibility are critical. DoD has taken some steps in improving the syntactic and semantic interoperability of some key data systems that collect acquisition data on large acquisition programs through open interfaces and data governance, but continued progress is needed (Anton et al. 2019). See Box 1.3 for the Federal Data Strategy approach to data accessibility and insight for DoD. For example, integrated data science is seen in DoD systems (e.g., in the integration of analytics in autonomous systems), but expanded integration in acquisition processes holds the potential for other systems as described in Chapter 4.

Adopting a data-informed approach to acquisition processes and management will require fundamental changes across DoD, and senior DoD leaders have an important role to play in making that vision a reality. The successful use and application of data are often facilitated by leaders who value data and encourage their use and who have secured some resources, established a routine workflow, and acquired personnel with the necessary skills for applying aspects of data science to acquisition, most commonly data analytics. The Secretary of Defense and other senior leaders can facilitate adoption by expressing an *expectation* that decisions are based on the most accurate information available. If these leaders insist on using data-driven processes to make these data-informed decisions, they must also demand that the data are high quality and as complete and accurate

> **BOX 1.3**
> **Data Accessibility in the Federal Data Strategy 2020 Action Plan**
>
> The Federal Data Strategy 2020 Action Plan identifies 40 practices to guide agencies, data users, and policy makers in improving data stewardship and creating value from data. One of the three broad groups of the practices is Governing, Managing, and Protecting Data. Of the 16 practices in this group, 9 address data access, sharing, and security.
>
> Readily available and accessible data, combined with the ability to share data with appropriate collaborators, enable:
>
> - Exploration, Analysis, and Reporting;
> - Collaboration and data integration to combat silos [Breaking down data silos] [The silo mentality];
> - Cross-program interaction through shared assets and common goals;
> - Reproducibility of findings;
> - Evidence based decision making;
> - Audit and accountability of decisions; and
> - Innovation.
>
> However, appropriate and principled data governance must be in place to provide policy and resources for data management, maintenance, use, and protection. To harness the power of data, it is important that key data assets be identified and made broadly accessible so that insights are not limited to a single analyst or organization. Collecting data is not sufficient; it must be curated, cleaned, formatted, stored, protected, and made easily accessible to support decision making.
>
> To make data more accessible across DoD, while reducing security risks, a clear process for requesting, granting, and revoking access (based on need-to-know, personnel clearance, data sensitivity, and any legal restrictions) must be in place. It should leverage DoD's capabilities in cybersecurity to guarantee the safety and integrity of the data.
>
> SOURCE: Department of Defense, 2020, *DoD Data Strategy*, September 30, https://media.defense.gov/2020/Oct/08/2002514180/-1/-1/0/DOD-DATA-STRATEGY.PDF.

as possible and provide the necessary resources. Such expectations and requirements will push these data-centric practices and the required use of data analytics and other aspects of data science down into other parts of the acquisition system. **By making decisions informed by data, leadership sets an example that can promote the cultural changes necessary to facilitate data-informed decisions across the entire department.**

The USD(A&S) has provided strong verbal support for data-informed decision making in defense acquisition—the following are just a few such examples. In her remarks on October 28, 2019, at the National Academies, Ms. Stacy Cummings, Principal Deputy Assistant Secretary of Defense for Acquisition, stated that the USD(A&S) prioritizes bringing innovation into the defense acquisition system through data, data-informed policy making, and an empowered workforce. Critical to this study, Ms. Cummings added that there is a need to transform training so that there is a baseline set of data-use skills throughout the defense acquisition workforce.

The USD(A&S)'s June 2020 memorandum, *Data Transparency to Enable Acquisition Pathways* (USD(A&S) 2020), highlights the value of providing decision makers access to comprehensive and accurate data. These statements in support of more data-informed decision making have been reinforced in the September 9, 2020, release of DoD Directive 5000.01 by the Deputy Secretary of Defense as well as the January 23, 2020, release of DoD Instruction 5000.02 by the USD(A&S).

In early 2020, USD(A&S) established the Adaptive Acquisition Framework, a set of acquisition pathways to help the workforce to tailor strategies to deliver better solutions faster (see DoD Instruction 5000.02, 2020). Meanwhile, through the Human Capital Initiatives,[2] USD(A&S) is initiating the Back-to-Basics initiative (USD(A&S) 2020), which is a major reform of the defense acquisition workforce management framework that will align with the Adaptive Acquisition Framework.

Given the recent adoption of the DoD Data Strategy 2020 Action Plan and the strong support from leadership, defense acquisition is well-positioned—from a policy perspective—to strengthen its data infrastructure and data-centric skills of its workforce.

Finding 1.1: By making decisions informed by data, leadership sets an example that can promote the cultural changes necessary to facilitate data-informed decisions across the entire department.

Recommendation 1.1: The Department of Defense (DoD) and Congressional leaders and stakeholders should promote the cultural changes necessary to facilitate data-informed decisions across the entire DoD by insisting that high-quality, complete, and accurate data and analysis be provided to inform their own decisions and those of their subordinates.

STUDY ORIGIN, OBJECTIVES, AND APPROACH

Over recent years, the National Academies of Sciences, Engineering, and Medicine has convened several activities to identify the challenges

---

[2] See https://www.hci.mil/.

of, and highlighting best practices in, postsecondary data science education (NASEM 2018, 2020a; NRC 2015). During discussions around these activities, representatives from DoD highlighted needs for strengthened appreciation and understanding of data science within the defense acquisition workforce, leading to discussions about creating a consensus study tailored to this context.

Around the same time, Congress passed the 2018 National Defense Authorization Act, directing DoD to establish a "set of activities that use data analysis, measurement, and other evaluation-related methods to improve acquisition program outcomes" (Section 913). These activities are to include the establishment of focused research and educational activities at the Defense Acquisition University[3] and appropriate private-sector academic institutions to support enhanced use of data management, data analytics, and other evaluation-related methods to improve acquisition outcomes.

At the request of USD(A&S), the National Academies convened a committee of diverse experts to help address the need for improved data use within the defense acquisition workforce. This study was conducted over 18 months, included a public workshop in April 2020 with a published proceedings (NASEM 2020b), and produced this consensus report. The complete description of the committee's charge can be seen in Box 1.4. A list of committee members along with their biographies is shown in Appendix F, and a list of meetings and information-gathering sessions held by the committee is shown in Appendix A.

Terminology—from both the DoD acquisition and data science communities—and a common understanding of terms are critical to addressing the Statement of Task. In particular, there are challenges in translations and contexts of terms and phrases; for example, "analysis" has different definitions in the two communities. Accordingly, throughout the study and within this report, the study committee sought to establish a common understanding of critical terms and be consistent with their use. Please see Appendix E for a glossary of these terms and phrases.

REPORT OUTLINE

This report is structured as follows. For short versions of background topics, Chapter 2 gives a brief summary of defense acquisition and its workforce and Chapter 3 discusses data science and the data life cycle. Chapter 4 explores the current uses of data in defense acquisition and identifies opportunities to exploit data and improve program and institutional performance. Chapter 5 identifies the portfolio of skills and organizational options for data science teams, and how supervisors with non-technical

---

[3] Recently renamed just "DAU."

> **BOX 1.4**
> **Statement of Task**
>
> The National Academies of Sciences, Engineering, and Medicine (the National Academies) will convene an ad hoc committee to execute a workshop and in-depth consensus study to identify relevant data science skills and capabilities necessary for the acquisitions workforce and develop a framework for training and educating acquisition professionals. Specific questions to be considered by the National Academies committee during the workshop and consensus study include:
>
> - How can data science improve acquisition processes and where are the opportunities to improve workforce ability to apply these methods?
> - What are the foundational understanding and skills that should be developed broadly in acquisition professionals, and what more advanced capabilities are relevant for specific job functions?
> - What are the characteristics and portfolio of skills of successful data science teams and how can supervisors with non-technical backgrounds effectively manage data science projects?
> - What data science training and education models exist in other government agencies and outside of government for employee training and up-skilling?
>
> The workshop will be recorded and webcast live, and a rapporteur will summarize the presentations and discussions in a "Proceedings of a Workshop." At the end of the study, the committee will produce a consensus report providing findings on how the Department of Defense can accelerate data analysis capabilities within the acquisition workforce.

backgrounds can effectively manage data science projects. Techniques and programs for upskilling the acquisition workforce are explored in Chapter 6 using best practices and lessons learned from academia, industry and other government agencies. Finally, this report's findings, conclusions, and recommendations are listed in Chapter 7.

REFERENCES

DoD (Department of Defense). 2019. "Recommendation 89: Direct DoD to consolidate or eliminate competing data architectures within the defense acquisition and financial system." Report of the Advisory Panel on Streamlining and Codifying Acquisition Regulations. Volume 3. January. https://discover.dtic.mil/wp-content/uploads/809-Panel-2019/Volume3/Recommendation_89.pdf.

DoD. 2020a. "DoD Data Strategy." September 30. https://media.defense.gov/2020/Oct/08/2002514180/-1/-1/0/DOD-DATA-STRATEGY.PDF.

DoD. 2020b. "DoD and OSD Component Heads." Washington Headquarters Services. November 23. https://www.esd.whs.mil/Portals/54/Documents/DD/iss_process/DoD_OSD_Component_Heads.pdf.

Miller, J. 2020. "The Big Data Challenge Getting Smaller for Army, Air Force as CDOs Mature," Federal News Network. January 24. https://federalnewsnetwork.com/ask-the-cio/2020/01/the-big-data-challenge-getting-smaller-for-army-air-force-as-cdos-mature/.

NASEM (National Academies of Sciences, Engineering, and Medicine). 2018. *Data Science for Undergraduates: Opportunities and Options*. Washington, DC: The National Academies Press. https://doi.org/10.17226/25104.

NASEM. 2020a. *Improving Defense Acquisition Workforce Capability in Data Use: Proceedings of a Workshop—in Brief*. Washington, DC: The National Academies Press. https://doi.org/10.17226/25922.

NASEM. 2020b. *Roundtable on Data Science Postsecondary Education: A Compilation of Meeting Highlights*. Washington, DC: The National Academies Press. https://doi.org/10.17226/25804.

NRC (National Research Council). 2015. *Training Students to Extract Value from Big Data: Summary of a Workshop*. Washington, DC: The National Academies Press. https://doi.org/10.17226/18981.

Sasala, T. 2020. "Perspectives from the Chief Data Officers." Presentation at the Workshop on Improving Defense Acquisition Workforce Capability. Virtual. April 14.

USD(A&S) (Under Secretary of Defense for Acquisition and Sustainment). 2020. "Data Transparency to Enable Acquisition Pathways." Department of Defense. https://www.acq.osd.mil/aap/assets/docs/USA000854-20%20Signed%20Memo.pdf.

Vidrine, E. 2020. "Perspectives from the Chief Data Officers." Presentation at the Workshop on Improving Defense Acquisition Workforce Capability. Virtual. April 14.

2

Defense Acquisition Process, Data, and Workforce: The Short Version

THE ACQUISITION PROCESS

To support the missions and needs of the U.S. military, the Department of Defense (DoD) needs a wide variety of platforms, weapons, supplies, and contractor services. Under the purview of the Under Secretary of Defense for Acquisition and Sustainment, the development and purchase of systems, services, and interdependent capabilities are governed by DoD's tailored acquisition policy, illustrated at a high level in Figure 2.1. The systems and capabilities are acquired via relevant pathways. Each pathway has specific instructions and supporting guidance and processes instructions that help govern and facilitate the way DoD develops and procures the associated capabilities. For example, a capability that is urgently needed by the operating forces usually follows a compressed development and production schedule of less than two years, but the authorities and definition of "urgent" is carefully controlled by Congress given the flexibilities they offer. In contrast, a major military capability is typically developed and procured over a longer period of time, often years with continued procurement and sustainment for decades. Aircraft, ships, submarines, spacecraft, and large ground vehicles fall into this category, but so do smaller vehicles or small unmanned systems. These capabilities are divided into subcategories depending on their cost. The higher the cost, the more senior the oversight and the more reporting that is required by Congress. Software systems follow another process and defense business systems yet another. These processes differ based on the nature of what is being acquired. For example, buying and upgrading software needs to follow a very different design and

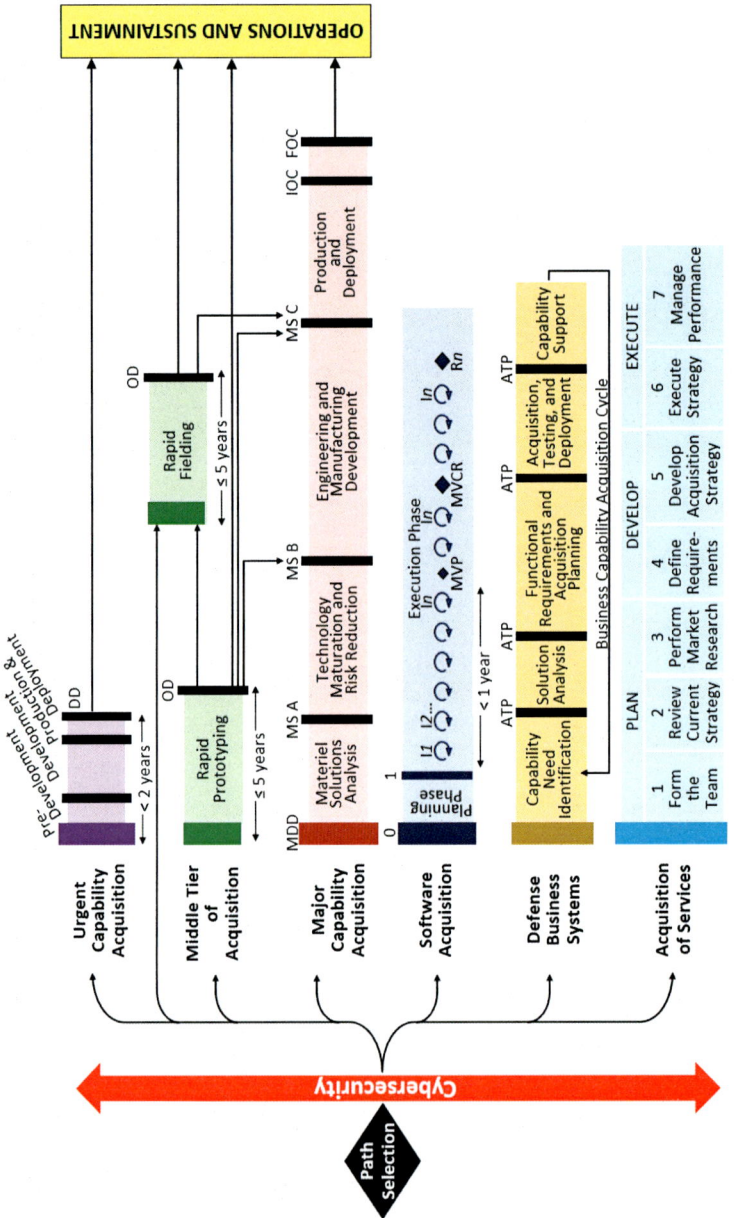

FIGURE 2.1 Adaptive Acquisition Framework. SOURCE: Department of Defense. "DoD Instruction 5000.02: Operation of the Adaptive Framework." 2020. https://www.esd.whs.mil/Portals/54/Documents/DD/issuances/dodi/500002p.pdf?ver=2019-05-01-151755-110.

procurement process than the process necessary to design and procure a new tactical aircraft or bomber with major physical elements. Finally, DoD acquires many contractor services, which follow yet another process.

ACQUISITION DATA USE AND CONTEXT

To acquire anything in DoD, authoritative determinations must be made for its necessity (through a requirements process) and funding (through a budget, authorization, and appropriation process). Each of the military departments has its own data systems to support the acquisition, requirements, and budget processes. While some data flows from the military departments into common DoD-wide data systems, there are still extensive data sets within each department using differing terminologies—even when referring to the same concept or piece of information—which magnifies the complexity of accessing and analyzing data across DoD.

The nature of what is being purchased governs the specificity and review process associated with the acquisition process and the tailoring of that process to the specific system in question. An expensive major weapon system (more than $3 billion in constant 2020 dollars) requires an operational requirement approved by senior military leaders and a budget allocation requested by the Secretary of Defense and the President for authorization and appropriation by Congress. Data and data analytics are extremely valuable in producing, for example, an accurate cost estimate when estimating the budget that will be required for procurement as well as when evaluating whether a postulated or prototyped weapon system will actually fulfill the operational need. Approval authority often is delegated to less senior officials as the cost of the program decreases; however, data analytics that support a needs assessment and a cost estimate are still needed at all levels.

At the other end of the spectrum, many non-developmental goods (such as parts or commercial supplies) and services are procured through purchasing contracts. These capabilities are also aligned with operational needs and may involve formal requirements and specific budgeting, but the oversight for underlying procurements may not be individually governed by the pathways outlined in Figure 2.1. Instead, a budget is allocated and the DoD official responsible for procuring the capability is expected to buy what is needed by getting a reasonable price. Again, data analytics are extremely valuable when assessing both the quantity of needed parts and supplies and the corresponding "fair" or "best value" price.

THE ACQUISITION WORKFORCE

The DoD acquisition workforce community is roughly 182,671 people strong as of the third quarter of FY 2020 (DoD 2020a) and a broad collection of disciplines that include program managers; engineers of nearly every discipline including test, manufacturing, and production specialists; financial and contracting professionals; and logisticians. About 10 percent are military and 90 percent are contracted support (DoD 2020; Schwartz et al. 2016).

Ensuring that DoD has a competent and trained acquisition workforce has been a focus and concern for decades. Dating back to at least the Packard Commission in 1986, there have been efforts to improve the management and training of the overall acquisition workforce. The Defense Acquisition Workforce Improvement Act of 1990 requires the Secretary of Defense to establish education and training standards, requirements, and courses for the civilian and military acquisition workforce. Those responsibilities are generally delegated to the USD(A&S). This Act also established the Defense Acquisition University (DAU), headquartered in Fort Belvoir, Virginia. In addition to DAU, members of the acquisition workforce may receive training at the military service academies—the Air Force Institute of Technology, the Naval Postgraduate School, and the National Defense University—and in public and private colleges and universities. While many members of the acquisition workforce receive education and training at these institutions, only DAU is dedicated to acquisition.

Because of its prominent role in training the acquisition workforce, committee members visited DAU to gain an appreciation for what it was teaching the DoD acquisition community with regard to data science. During its visit in January 2020, the committee learned that DAU offered few dedicated courses in data science, but that several courses included aspects of data analytics. During the same visit, discussions with members of a hands-on course for program managers disclosed that data use and data analytics issues were not consistently embedded in practical exercises and that some class members felt unprepared to find or use data in decision making. To facilitate application, most data use and data analytics training are embedded with acquisition disciplines rather than taught as a separate skill. These observations are consistent with recent analyses by Anton et al. (2019). However, in October 2020, DAU released "CENG 002: Data Analytics for DoD Acquisition Managers Credential." According to the DAU catalog, CENG 002 is a 32-hour course introducing data science to mid-level acquisition workforce managers in all career fields seeking to use data to inform decisions. Its timely release is indicative of the rapidly changing data science environment and DoD's sense of urgency in upskilling its workforce to better understand and embrace the possibilities of data. Other approaches to data science and data analytics training for the acquisition workforce are covered in Chapter 6 of this report.

In summary, the DoD acquisition processes are as varied as the types of products and services required to support the military departments. Data are complex and often characterized differently within different programs or systems. Finally, members of the acquisition workforce are very skilled but—reflecting the many heterogeneous processes they support—have diverse sets of skills, roles, and experiences. Given the diversity and complexity of acquisition in DoD, a single, all-encompassing solution for improving workforce capability in data use is unlikely. However, a framework for training and educating the workforce—customizable to processes, skillsets, and roles—may enable enhanced data use.

REFERENCES

Anton, P.S., M. McKernan, K. Munson, J.G. Kallimani, A. Levedahl, I. Blickstein, J.A. Drezner, S. Newberry. 2019. *Assessing Department of Defense Use of Data Analytics and Enabling Data Management to Improve Acquisition Outcomes*. RR-3136-OSD. Santa Monica, CA: RAND Corporation. https://www.rand.org/pubs/research_reports/RR3136.html.

DoD (Department of Defense). 2020a. "Defense Acquisition Workforce Key Information: Overall." FY20Q3. June 30. https://www.hci.mil/docs/Workforce_Metrics/FY20Q3/FY20(Q3)OVERALLDefenseAcquisitionWorkforce(DAW)InformationSummary.pptx.

DoD. 2020b. "OSD/JS Privacy Program." https://www.esd.whs.mil/Portals/54/146/Privacy/Frequently%20Asked%20Questions%20(2017%20update).pdf.

Schwartz, M., K.A. Francis, and C.V. O'Connor. 2016. "The Department of Defense Acquisition Workforce: Background, Analysis, and Questions for Congress." Congress Research Service. July 29. https://fas.org/sgp/crs/natsec/R44578.pdf.

3

Data Science and the Data Life Cycle: The Short Version

WHAT IS DATA SCIENCE AND WHO DOES IT?

Everyone consumes, processes, and interacts with data every day, and everyone makes decisions based, in part, on data. For example, we use weather forecasts to make plans for a day's activities. We purchase groceries based on combining our past meal history and future dining plans. We scroll through and choose from a list of recommended news articles selected for us by algorithms that combine our interests with current events. We make decisions that impact our national security by integrating, visualizing, and analyzing different data sources. These examples show the wide range of areas in which data are routinely used—and in which the more systematic use of data science might yield better decisions.

In this section, the committee more clearly defines data science by describing it as a multi-phase process—facilitated by people—that extracts value from data to answer posed questions. Understanding this definition and phases within it are fundamental to incorporating data science into defense acquisition. In Box 3.1, the committee outlines three key features of data science that shape this chapter and summarize the basics for the defense acquisition community.

DATA SCIENCE IS COLLABORATIVE AND CYCLICAL

Many of the early definitions of data science focused on identifying static collections of skills necessary for workforce members to have the title of "data scientist." Typically visualized via Venn diagrams, these definitions

> **BOX 3.1**
> **Data Science and the Data Life Cycle:**
> **Key Takeaways for Defense Acquisition**
>
> 1. Extracting decision-quality information from data requires a multidisciplinary team. To do data science, leaders in defense acquisition are not looking for utopian "unicorn" individuals with expertise in all things data; they need a team.
> 2. Data science is a cyclical process with many phases. Like acquisition life cycles, different skills are more important at different phases of the data life cycle.
> 3. Much like acquisition is infused with concerns such as cyber protection, physical security, and intellectual property, data science is cocooned in protections such as data ethics, data security, and data privacy.

highlighted overlapping academic disciplines (including computer science, engineering, mathematics, statistics, and social sciences), which emphasized the cross-disciplinary nature of data science at the central intersection of the diagram. Figure 3.1 below shows an example of these types of data science Venn diagrams (Geringer 2014); although in this case the word "unicorn" is in the center, indicating that finding that one single person with mastery of all related disciplines is akin to searching for a mythical creature. The Venn Diagram framework, while a useful visual from a traditional disciplinary point of view, does not capture data science *in practice*.

Noting a tremendous demand for data scientists in the United States and some uncertainty for their education, the National Academies of Sciences, Engineering, and Medicine undertook an earlier consensus study titled *Data Science for Undergraduates* (NASEM 2018). This study included characterizing data science as being centered on "the notion of multidisciplinary and interdisciplinary approaches to extracting knowledge or insights from large quantities of complex data for use in a broad range of applications" (p. 13). Further, "data science is not just the practice of analyzing a certain data set about a particular question. It often results in the creation of processes that continuously take in new data, often from many sources, and generate refined distillations of that data, which in turn become sources for new inquiries, questions, and analyses" (p. 14). In keeping with this characterization, data science is more commonly viewed now as a *process* or *workflow* in which real problems are solved with real data through an often-cyclical set of phases requiring different skillsets.

In her 2020 *Harvard Data Science Review* article, Jeannette Wing wrote, "Data science is the study of extracting value from data. 'Value' is subject to the interpretation by the end user and 'extracting' represents the work done in all phases of the data life cycle" (Wing 2020). The data life

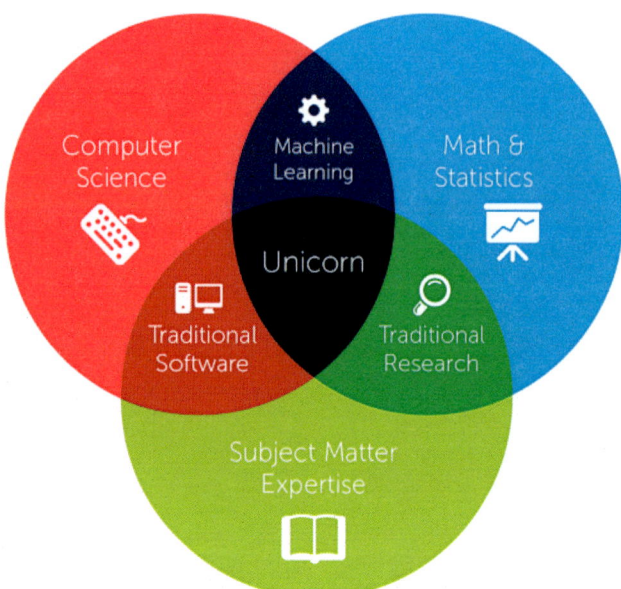

FIGURE 3.1 Geringer data science Venn diagram. SOURCE: Copyright © 2014 by Steven Geringer, Raleigh, NC.

cycle included in the article appears in Figure 3.2; it shows a workflow that includes phases: generation, collection, processing, storage, management, analysis, visualization, and interpretation. Special emphasis is placed on the importance of security/privacy and ethical concerns in all phases. The data life cycle also begins with the integration of disparate sources of information and data, ending with dissemination, consumption, and adoption by stakeholders, which are phases crucial to optimizing the value of data.

This notion of a data life cycle has also been accepted beyond the data science community, including the federal government. Figure 3.3, featured in the Federal Data Strategy: Improving Agency Data Skills Playbook and adopted from the National Institute of Standards and Technology (NIST), depicts a comprehensive and well-managed data life cycle with similar steps to those in Figure 3.2. Note that in the below diagram, as is noted in the 2018 National Academies of Sciences, Engineering, and Medicine consensus study on data science for undergraduates, after assessment, implementation, and feedback from data consumers, the cycle returns to the beginning where disseminated results inform the development of new inquiries for data sources.

Cyclical workflow diagrams are seen within the Department of Defense (DoD) that include most aspects of the data life cycle, albeit labeled primar-

DATA SCIENCE AND THE DATA LIFE CYCLE 27

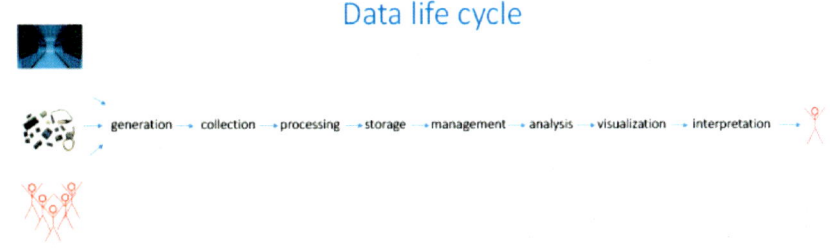

FIGURE 3.2 Wing's data life cycle. SOURCE: J.M. Wing, 2020, "Ten Research Challenge Areas in Data Science," *Harvard Data Science Review*, https://doi.org/10.1162/99608f92.c6577b1f. Copyrighted Jeanette M. Wing (2017, 2019).

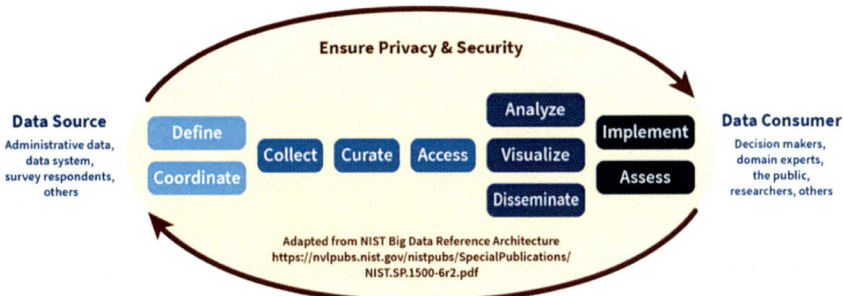

FIGURE 3.3 NIST data life cycle.

ily as "data analytics"; for example, see Figure 3.4 for a cycle that describes the Air Force "data analytics ecosystem."

While these diagrams capture the multi-phase workflows associated with extracting value from data to solve problems, they have less emphasis on the initial questions and stakeholders, (possibly inadvertently) downplaying the central role that people play in a successful data science workflow (Marshall and Geier 2019). They also do not capture the iterative process that commonly occurs as people return to previous phases to correct errors or reexamine strategy—for example, moving back and forth between analysis and visualization as statistical modeling is fine-tuned. A data science workflow, while tending toward a direction, rarely travels in a straight line.

In her remarks at the National Academies on January 31, 2020, Professor Sallie Ann Keller shared a data science framework that addresses many shortcomings of these diagrams. This framework, used at the Biocomplexity Institute at the University of Virginia, emphasizes that the multi-phase process of data sciences includes problem identification (via questions and

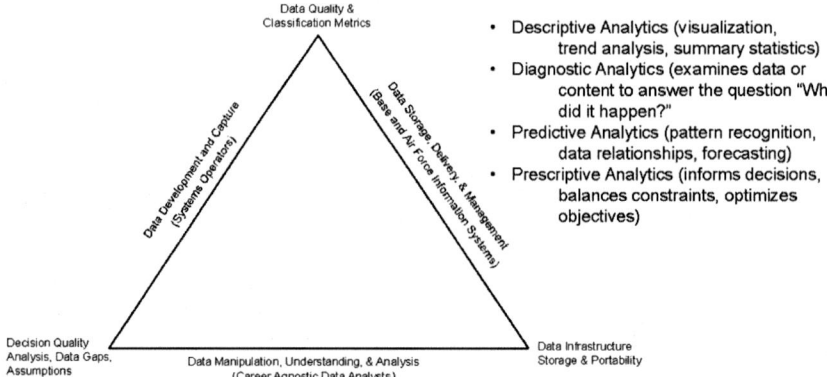

FIGURE 3.4 Air Force data analytics ecosystem.

working hypotheses), communication, dissemination, and data wrangling (Keller et al. 2020). Keller concluded her remarks by saying "the data science framework enables creation of repeatable and measurable processes for the use of and repurposing of all data sources."

THE DATA LIFE CYCLE AND ITS PHASES

Building upon these diagrams and for use in this report, the committee uses a workflow that (1) incorporates questions, (2) is more cyclical (as opposed to linear) and iterative, (3) communicates the need for small loops, (4) enables multiple entry points, and (5) emphasizes the role of people. With a slight modification to Wing's definition of data science, the committee will use the definition in Box 3.2 for the remainder of the report.

Figure 3.5 depicts a bi-directional diagram with many people (including stakeholders) as a central focus along with an emphasis on data privacy, ethics, and security. Critically, data science is facilitated by people who make decisions about when to move forward and/or backward and are responsible for minimizing harm. It is important to note that the people at

---

BOX 3.2
Definition of Data Science

Data Science is the science and technology of extracting value from data and is characterized by the data life cycle (as shown in Figure 3.5).

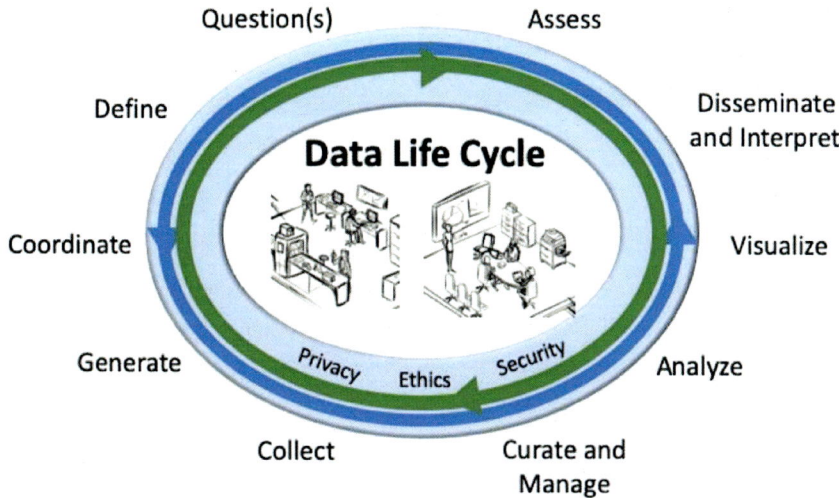

FIGURE 3.5 Data life cycle.

the center of the data life cycle can have different roles in facilitating data science. This feature will be discussed in Chapter 5.

The ideal entry point is the *Question(s)* phase; the process then moves along the blue arrow, returning to previous phases as needed along the green arrow. Once results are disseminated and interpreted, the assessment phase allows stakeholders an opportunity to refine their questions, beginning the cycle again. To better characterize how each phase in the data life cycle above contributes to data-informed decision making, the committee further describes the phases and corresponding questions and actions as follows:

- *Question*—Data science extracts value from data to help answer a posed question or inform a decision. Questions are often developed from engagement with stakeholders; decisions can be path-critical turning points where leadership may need to determine whether or not to move forward with a new product, process, or program.
- *Define*—Is the question or decision well-posed? Is it "answerable"? Which data are required to inform the decision or answer the question? Do the data exist already, or must they be generated? What data quality is required? Data quality includes considerations of accuracy, precision, completeness, authoritativeness, consistency, reliability, relevance, and timeliness.
- *Coordinate*—Data must be accessible and available for analysis. Can current resources and infrastructure provide the required data? If not, what else is required?

- *Generate*—There are many mechanisms for generating data: for example, some data are operational (e.g., from business processes or front-line operations), other data come from sensors, and yet other data originate from surveys. In the context of the question or decision, what data exist or should be created?
- *Collect*—Not all generated data should be collected and stored for future use. Which data should we collect and store? How should the collected data be organized? What are the limitations of the data? How were the data created and sampled, and what impact does that have on its relevance for the question or decision?
- *Curate and Manage*—In order to provide value, data must be organized, refined, and maintained with sufficient quality to support decisions and answer questions. Data curation and management have multiple sub-tasks:
  — *Process*—Clean, wrangle, format, compress, encrypt/decrypt, and authenticate data.
  — *Store*—Identify the appropriate data storage system and create appropriate metadata to maximize the ability to access and modify the data for subsequent analysis while ensuring proper data security.
  — *Integrate, Fuse, Link*—If relevant to underlying processes and questions, build metadata hierarchies that allow for linking data sets through common IDs.
  — *Access*—Implement appropriate data access methods, including system interoperability and other infrastructure to share data across pieces of the enterprise as needed.
- *Analyze*—Data analysis encompasses the use of data and tools to generate insights through statistical modeling, machine learning algorithms, visualization, and human examination. These activities include descriptive statistics and predictive analysis.
- *Visualize*—One of the most effective ways to present analysis results in clear, simple, interpretable forms is to use visualization techniques such as histograms, scatterplots, interactive charts, time-dependent graphs, and heat maps. Visualization methods are also commonly used prior to and during the analysis phase to help inform next steps.
- *Disseminate/Interpret*—Explain what the analysis and results mean in a way that is interpretable and appropriate to the decision maker and stakeholders.
- *Assess*—Continuously monitor and improve all processes in the data life cycle. Use the current analysis and results to refine and develop subsequent questions and decisions.

DATA ETHICS, PRIVACY, AND SECURITY

Elements of data ethics, privacy, and security are incorporated throughout the data life cycle. The American Statistical Association defines good statistical (and by extension, data science) practice as "fundamentally based on transparent assumptions, reproducible results, and valid interpretations." This includes using methodology and data that are relevant and appropriate; being transparent about any known or suspected limitations or biases in the data that may affect the reliability of the analysis; protecting the interests of the respondents whose data are considered; and considering the entire range of explanations for observed phenomena (ASA 2018). Addressing threats to privacy—increased by broader data access, machine learning, and artificial intelligence—is an ongoing challenge for policy makers and practitioners. In "Data, Privacy, and the Greater Good," Eric Horvitz and Deirdre Mulligan note that "machine learning can be used to draw powerful and compromising inferences from self-disclosed, seemingly benign data or readily observed behavior. These inferences can undermine a basic goal of many privacy laws—to allow individuals to control who knows what about them." They also note that the White House and the Federal Trade Commission (FTC) have, in the past, sought to protect "privacy, regulate harmful uses of information, and increase transparency" (Horvitz and Mulligan 2015). See Box 3.3. In many applications, the appropriate level of security for data through their life cycle is critical. Data can be manipulated, stolen, or destroyed. Resulting risks vary depending on data types and sources; extreme cases include intellectual property theft and threats to national security.

In the opinion of this committee, the DoD acquisition workforce has been leveraging data science, even if not identified as such. Nevertheless, in Chapter 4, the committee identifies some opportunities for improved data use and describes how the data life cycle can be incorporated into common acquisition functions.

> **BOX 3.3**
> **Legal and Ethical Considerations in the Data Life Cycle**
>
> Privacy and other legal and ethical considerations affect what data are collected, who can access them, what analyses are done, and even what kinds of questions are addressed. Thus, data scientists, leaders, and stakeholders need some level of familiarity and experience in making legally informed and ethical decisions about data, along with access to resources that they can consult for deep knowledge and guidance on these matters.
>
> There are many instances when statutes, regulations, or policies do not cover the specific details of a case or are not well developed. In these cases, personnel may need to be prepared to analyze these cases and make careful decisions about what they should do because policy and guidelines are not mature. For example, caution may be warranted in collecting data for developing artificial intelligence analytic engines that make military decisions, such as whether to impose lethal force. Other examples may be systems that apply bias in social settings that may not yet be in line with evolving social norms (e.g., when data science may lead to racial profiling that may not be prohibited or may have conflicting policies).
>
> Not all ethical and moral considerations relate to *limitations* on what data science can be performed. In some cases, data science can provide objective insights to *inform* policy making or monitor situations to ensure that ethical and moral considerations are being addressed. As with all data science, these insights cannot be compelling unless the data are sufficient, the analysis is rigorous, the methods are not overly complex, and the results are understandable.

REFERENCES

ASA (American Statistical Association). 2018. *Ethical Guidelines for Statistical Practice.* https://www.amstat.org.

Geringer, S. 2014. "Data Science Venn Diagram v2.0." Steve's Machine Learning Blog (Blog). January 6. http://www.anlytcs.com/2014/01/data-science-venn-diagram-v20.html.

Horvitz, E., and D. Mulligan. 2015. Data, privacy, and the greater good. *Science* 349(6245): 253-255.

Keller, S.A., S.S. Shipp, A.D. Schroeder, and G. Korkmaz. 2020. "Doing Data Science: A Framework and Case Study," HDSR, February 21. https://hdsr.mitpress.mit.edu/pub/hnptx6lq/release/6https://hdsr.mitpress.mit.edu/pub/hnptx6lq/release/6.

Marshall, B., and S. Geier. 2019. "Targeted Curricular Innovations in Data Science." 2019 IEEE Frontiers in Education Conference (FIE). https://ieeexplore.ieee.org/abstract/document/9028491/https://ieeexplore.ieee.org/abstract/document/9028491/.

NASEM (National Academies of Sciences, Engineering, and Medicine). 2018. *Data Science for Undergraduates: Opportunities and Options.* Washington, DC: The National Academies Press. https://doi.org/10.17226/25104.

Wing, J.M. 2020. "Ten Research Challenge Areas in Data Science." *Harvard Data Science Review.* https://doi.org/10.1162/99608f92.c6577b1f.

4

Data Science in DoD Acquisition

The use of data to support decision making is not new to the acquisition community and has facilitated critical program support and even informed acquisition strategies for decades—see Appendix C. However, an increasing emphasis on data collection and management, together with use of data tools and associated visualization and analysis techniques, offer the Department of Defense (DoD) an opportunity to create, tap into existing, enhance, or even institutionalize what might have been an ad hoc process into a cohesive, consistent data strategy across acquisition.

This chapter discusses several instances where the application of improved data use could augment the success of the acquisition programs, followed by a brief reminder of the breadth of acquisition functions engaged in data-informed decision-making. Finally, for several of these acquisition functions, this chapter explores data-informed decision opportunities and identifies accompanying phases in the data life cycle needed to enable or enhance those decisions.

OPPORTUNITIES FOR IMPROVED DATA USE IN DEFENSE ACQUISITION

Frank Kendall, the Under Secretary of Defense for Acquisition, Technology and Logistics from 2012 to 2017, was an avid data-informed decision maker. In fact, one of his Better Buying Power principles was that "Data should drive policy"; emphasizing this point was a sign posted outside his office door that read, "In God We Trust; All Others Must Bring Data" (Kendall 2016).

While defense acquisition data is collected in repositories such as the Office of the Secretary of Defense (OSD) Cost Assessment and Program Evaluation's (CAPE's) Cost Assessment Data Enterprise (CADE) system and reports such as Selected Acquisition Reports and the Defense Contract Management Agency's Program Assessment Reports have been available as data analytics resources for many years, the committee sought insights regarding access to defense acquisition data in testimony from previous leaders in the DoD acquisition community. In each instance, missed opportunities were identified for improved use arising from a lack of enterprise-wide data collecting, managing, and sharing practices.

Heidi Shyu, a former Assistant Secretary of the Army for Acquisition, Logistics, and Technology (ASA(ALT)), noted several areas where improved data science and data analytics might have been applied to enhance DoD acquisition programs and processes. Ms. Shyu acknowledged that data and information were available to inform her decisions, but she often found that the data were dispersed across the organization such that collection and analysis would be a tedious months-long process when data-informed decisions were needed daily. Her assessment was that DoD lacks the capability to easily share data across program offices, decreasing visibility across programs and enabling perpetuation of "siloed" programs.

She noted that a common data sharing infrastructure and common dashboard for accessing said data is an opportunity for enhancing DoD acquisition programs. Ms. Shyu cited a number of examples where centralized defense data would have enhanced her ability to make better data-informed decisions for the Army. These opportunities include:

- Budget management by managing multi-year budgets across program offices;
- Program management by enhancing cost, schedule, and performance management, risk assessments, milestone delivery, and program reviews;
- Supply-chain analysis and maintenance by identifying critical companies and potential vulnerabilities;
- Business intelligence by monitoring companies progress and financials;
- Personnel management by tracking personnel roles and skillsets, including past programs of personnel and performance metrics and expertise management; and
- Data management by interfacing across multiple disparate databases with substantial legacy data to maintain and process.

Dr. J. Michael Gilmore, a former Director of Operational Test and Evaluation (DOT&E) and a former Assistant Director for National Security at the Congressional Budget Office (CBO), spoke with the committee on the challenges of managing data in acquisition, incentivizing the acquisition community to make better use of data science, and the need for leadership. Dr. Gilmore reinforced the committee's findings regarding common data challenges with the following observations: Data are not monolithic, and there are challenges facing collection and collation of data, including costs, access, and lack of a central archive to house the data.

Even when leadership has created a culture where data are well collected, well curated, centrally available, shared, and analyzed, the committee found that decisions need to be data-informed, not blindly data-driven. For example, a report on *Assessing Department of Defense Use of Data Analytics and Enabling Data Management to Improve Acquisition Outcomes* by Anton et al. (2019) assessed several large acquisitions programs that had overrun their cost targets. In some cases, DoD had determined that decision makers had failed to act on available data information (DoD 2016b, p. 28). Further analysis, however, identified other considerations (such as pressing military needs, risks, budgets, and political considerations) that played into the acquisition decision process. While acquisition decisions involve contextual considerations, the quality of the data and associated analyses that reach the decision maker is still an important input to acquisition decision making. The better the data and the more competent the workforce, the better the analyses and the better informed the decision maker will be.

DEFENSE ACQUISITION FUNCTIONS AND THE DATA LIFE CYCLE

One of the challenges in identifying opportunities for applying data science to defense acquisition is that the applications must be deeply embedded in the acquisition functions and processes—and rightly so (Anton et al. 2019). Thus, it is useful to have at least a top-level understanding of the range of acquisition functions. Box 4.1 provides a list of those functions, showing the multi-faceted activities involved in defense acquisition (Anton et al. 2019). Further details on how the range of acquisition functions and processes align with analyses in support of decisions are illustrated in Figure B.5 of Appendix B.

By combining the list in Box 4.1 with the data life cycle depicted in Figure 3.5, one can begin to see potential for use of data science on specific acquisition functions; below, the committee describes several opportunities for improvement.

> **BOX 4.1**
> **Primary Acquisition Functions**
>
> 1. Program Management/Manager
>   - Business case and economic analysis
>   - Affordability analysis
>   - Acquisition strategy
>   - Risk management
>   - Technical maturity
>   - Personnel and team management
>   - Business and marketing practices
>   - Configuration management
>
> 2. Research and Development (R&D)
>
> 3. Engineering
>   - Systems engineering
>   - Facilities engineering
>   - Software/IT
>
> 4. Intelligence and Security
>   - Cybersecurity
>   - Program Protection
>
> 5. Test and Evaluation (T&E)
>   - Developmental T&E
>   - Operational T&E
>
> 6. Production, Quality, and Manufacturing (PQM)
>
> 7. System and Operational Issues
>   - Spectrum (frequency allocation, emissions, etc.)
>   - Environmental
>   - Energy
>
> 8. Product Support, Logistics, and Sustainment

Contracting and Supply-Chain Cost

Government contracting officers are obligated to ensure that the government is paying a fair and reasonable price, often using cost insights to establish the final contract price. Supply chain costs are built into the prime contractor's price, and the supply chain feeds the development, production and sustainment pipeline for a system's lifetime. Often, each contract is siloed within each military service. The contracting officer (in conjunction

9. Financial Management

10. Cost Estimating

11. Auditing

12. Contract Administration
    - Contracting actions
    - Contracting strategy
    - Contract peer review
    - Acceptance of deliverables

13. Purchasing

14. Industrial Base and Supply-Chain Management

15. Infrastructure and Property Management

16. Manpower Planning and Human Systems Integration

17. Training and Education
    - Training and education for government execution
    - Training and education for acquired systems

18. Disposal

Acquisition Interface Functions

19. Requirements: receive, inform, and fulfill

20. Acquisition Intelligence: request, receive, and respond

21. Legal Counsel: request and act upon

SOURCE: Anton et al. (2019).

---

with the program manager, if there is one) is responsible for determining the contractor's past cost, schedule, and performance history to understand past performance, independently estimate costs for the current solicitation, and assess the reputation and reliability of suppliers. Contracting officers also monitor the status of current contracts, ensure deliverables are received, authorize payments, and report on contractor performance.

In the future, the data life cycle, as applied to this contracting function, suggests that DoD contracting systems could *collect*, *curate*, and *share* aggregated historical contract data for vendors across programs, even across the military services. This could allow, for example, the identification of subcontract management issues with problematic supply chain vendors less likely to deliver on time due to internal or external factors and use data *analysis* to predict which contracts or programs are more likely to be affected.

Sustainment of Acquired Systems

Roughly 70–80 percent of a DoD acquisition program's total cost is in the sustainment of the acquired system. As systems age, the costly and time-consuming effort of replacing obsolete parts becomes an increased concern, particularly as today's systems increasingly rely on electronics.

In the past decade, data science—together with computer science and information technology—has been used to create "digital twin" parts to model the lifetimes and replacements of the thousands of parts comprising each system (Raghunathan 2010). More recently, evolving digital engineering capabilities demonstrated in the Air Force's e-T7A Red Hawk (USAF 2021) will likely improve predictability in maintenance and parts service life of acquired systems. Investing and increasing in the *generation*, *collection*, *storage*, and subsequent *modeling* of data for designs, parts, and processes will enable DoD to rapidly respond to changing threats with real-time testing of new components in any system, and ultimately make better informed decisions regarding the sustainment, improvement, and retirement of military systems over time (Mizokami 2020).

Business Case and Economic Analysis

For any new acquisition program, DoD needs requirements, a business case, and economic analysis to justify the expenditure and acquisition approach. This includes an assessment of the return-on-investment and economic trends underlying the capability provided by the program. An investment in a new technology that is expensive but that should provide a significant military advantage would provide a justifiable return-on-investment. If a capability was projected to be at the early height of its cost-maturity curve and it was not a vital capability for military operations, the economic analysis would allow decision makers to decide whether or not to delay the acquisition to take advantage of declining prices.

These tradeoffs are currently conducted using whatever information is readily available about threat capabilities, technology advancements, and the cost of new technologies. The data life cycle suggests that these data could be *defined* more broadly, *collected* and *curated* across all of DoD, and made available to the acquisition workforce in the military departments and OSD. With better data, more appropriate data analytics could be applied and the results *visualized* in a way that would show trends or tipping points that would be helpful in designing acquisition programs to maximize benefits and minimize costs.

Cybersecurity

Recently, DoD has created a Cyber Maturity Model Certification[1] to assess and ensure the cybersecurity posture and readiness of contractors in the defense industrial base. As DoD attempts to increase cybersecurity and prevent the loss of intellectual property and weapon system designs, tracking and sharing the cybersecurity maturity of different contractors will be very helpful to program managers in selecting among bidders on source selections. By *collecting, curating,* and *sharing* the cybersecurity maturity level of different contractors over time, *data analyses* could be designed to identify cybersecurity weaknesses. Trends and strengths *assessments* could be conducted, and the results could be used to inform acquisition and contract decisions.

Cost Estimating

The ability to accurately estimate the cost of a new program is necessary to support investment decisions and tradeoffs within a budget and to set reasonable baselines to assess cost performance over time. As discussed earlier, CADE contains an established historical database that provides detailed *collected* cost and schedule data for a variety of prior systems. This database, which is *shared* across all of DoD, supports data analytics by helping cost analysts estimate the costs of new programs based on extrapolations from historical data. CADE is an existing widely used example of the value of using the data life cycle.

Acquisition and Contract-Management Strategies

As discussed earlier, Heidi Shyu identified multiple areas where improved data *collection, sharing,* and *analyses* would have been helpful to her as she worked to develop successful acquisition strategies. She suggested many areas where she and other decision makers would have benefited from better data. By aligning more fully with the data life cycle the information she needed would be available to future acquisition leaders. For example, DoD could *collect, curate,* and *share* information about:

- The personnel roles and skillsets resident in various contractors,

---

[1] See https://www.acq.osd.mil/cmmc/.

- Companies that provide critical capabilities but that may be experiencing financial stress, and
- General business intelligence about companies that provide needed capabilities procurement.

With this information, acquisition and contract-management strategies could be better designed by the acquisition leadership. By making these data *accessible* across DoD, use of the data life cycle would allow DoD to avoid poor performers and ensure critical capabilities were available. It could also identify duplication and opportunities for creating larger, cross-DoD contracts that might lead to cost savings. For example, earlier in the data life cycle, DoD could search for already existing data sources, internal or external, that could be used with minimal effort.

In addition to these case studies above, Table D.2 provides some additional but brief examples of acquisition-related functions and decisions using multiple phases of the data life cycle.

Finding 4.1: Data science has improved acquisitions processes by enhancing program and contracting analysis, tracking cost and program performance, enabling analysis of alternatives, and informing decision making.

Finding 4.2: There are several current shortcomings in the front end of data science processes in defense acquisition—namely in data collection and curation.

Finding 4.3: Data silos, wherein data are not shared across DoD, are a common obstacle preventing the utilization of the full potential of the data life cycle.

Conclusion 4.1: With support from defense acquisition leadership, improved data-sharing policies and support for front-end phases of the data life cycle are major short-term opportunities to improve acquisition processes.

Conclusion 4.2: A common data infrastructure and platforms, personnel training and tools, and access to DoD-wide data are major long-term data science investments that will have long-term positive impacts on acquisition processes.

Conclusion 4.3: Improved data use and the data life cycle could be applied to urgent, high-profile challenges in defense acquisition. Doing so would serve as exemplars of the value of data use and data-informed decision-making in acquisition, serving as a catalyst for further change.

Recommendation 4.1: The Department of Defense should continue to seek improvements in defense acquisition through the increased application of data science, including addressing shortfalls in data collection, curation, management, and sharing.

REFERENCES

Anton, P.S., M. McKernan, K. Munson, J.G. Kallimani, A. Levedahl, I. Blickstein, J.A. Drezner, S. Newberry. 2019. *Assessing Department of Defense Use of Data Analytics and Enabling Data Management to Improve Acquisition Outcomes.* RR-3136-OSD, August. Santa Monica, CA: RAND Corporation. https://www.rand.org/pubs/research_reports/RR3136.html.

Kendall, F. 2016. "Better Buying Power Principles: What Are They?" Defense AT&L. January–February. https://www.dau.edu/library/defense-atl/DATLFiles/Jan-Feb2016/Kendall.pdf.

Mizokami, K. 2020. "The Air Force Debuts a New 'e' Aircraft Designation." *Popular Mechanics.* https://www.popularmechanics.com/military/aviation/a34043731/air-force-new-designation-e-series-aircraft/.

Raghunathan, V. 2019. "Digital Twins vs Simulation: Three Key Differences." *Entrepreneur.* https://www.entrepreneur.com/article/333645.

USAF (U.S. Air Force). 2021. "Air Force Acquisition Executive Order Unveils Next e-Plane, Publishes Digital Engineering Guidebook." January 29. https://www.af.mil/News/Article-Display/Article/2476500/air-force-acquisition-executive-unveils-next-e-plane-publishes-digital-engineer/.

5

Data Life Cycle Mindset, Skillset, and Toolset: Roles and Teams

The Department of Defense (DoD) acquisition community has a history of incorporating data analytics but recognizes the further potential in today's rapidly evolving data science environment. In Chapter 3, the committee introduced the data life cycle and how its full utilization would better prepare the defense acquisition workforce to extract value from data. While not all inclusive, Chapter 4 identified a number of data-rich opportunities still available to further inform decisions in the acquisition community.

Extracting value from data requires a collective data life cycle mindset, skillset, and toolset. In this chapter, the committee explores the constantly evolving ways industry, government, and academia are shaping the mindset, skillset, and toolsets of its employees and data science teams as well as how these best practices and trends yield opportunities for defense acquisition.

MINDSET

Businesses are adopting a new collective mindset that recognizes the inherent value of data. According to the NewVantage 2019 Big Data and Artificial Intelligence (AI) Executive Survey, "92% of the respondents are increasing their pace of investment in big data and AI." Of the 65 leading finance, healthcare, and manufacturing firms that responded, 88 percent felt a greater sense of urgency to invest in big data and AI, and 75 percent attributed their urgency to fear of disruption by new entrants in their marketplace (NewVantage 2019). These responses represent a growing sentiment in industry that every company is (Yueh and Bean 2018; Orad 2020), or will eventually be, a "data company" where data will be a central elements of the way they do business (Orad 2020).

Similar to industry, successful U.S. government-affiliated efforts have found that a strong, visible leadership commitment has begun to help overcome institutional inertia regarding new data tools and applications. In remarks given at the National Academies of Sciences, Engineering, and Medicine in March 2020, Sezin Palmer articulated challenges and lessons learned as a Johns Hopkins Applied Physics Lab data science team created a Precision Medicine Analytics Platform. The new platform promised to revolutionize patient diagnosis, prognosis, and treatment; yet, along with hurdles in technology, data quality, data privacy and security, there were cultural barriers. Per Palmer, visible senior leadership buy-in and commitment from inception through implementation of this innovative data tool were required to help overcome organizational challenges.

In 2018, Michael Conlin became DoD's first-ever chief data officer (CDO) followed by CDOs in place in each military department by Fall 2019. With a focus on data and a sense of urgency given the near-peer threats to U.S. national security, Mr. Conlin declared in his April 2020 presentation to the committee that he wanted DoD to move more rapidly in enabling digital operations and decision making, adopting a collective data mindset with the expectation that every DoD employee eventually be digitally savvy (Conlin 2020).[1] In addition, and as was noted in Chapter 1, DoD released its new data strategy in October 2020 with an emphasis on decision making at the senior level.

In 2019, Anton et al. found that "[s]ome of the biggest barriers to expanding and refining the use of data analytics in the acquisition sphere include the lack of data sharing because of cultural, security, and micromanagement concerns; inconsistent data access across DoD and for FFRDCs [federally funded research and development centers] and support contractors; and difficulty installing modern analytic software because of security concerns." In addition, they note that "[l]ong-term investments and strategic planning are needed—both for data governance and for analytic capabilities—as well as concerted efforts by Congress and DoD to address the culture of not sharing data" (Anton et al. 2019a).

To remedy the issues identified by Anton et al., in the 2020 Defense bill Congress mandated that the DoD CDO "shall have access to all Department of Defense data, including data in connection with warfighting mis-

---

[1] There is no single definition of what it means to be digitally savvy, but "savviness" relates to practical knowledge and ability (Oxford English Dictionary, https://dictionary.cambridge.org/us/dictionary/english/savvy). There are digital literacy frameworks that, like those for data literacy, help convey the essential skills in understanding how to use and apply digital technologies (see, e.g., Van Deursen et al. 2014; MediaSmarts 2016; Hadziristic 2017; Huynh and Do 2017; Kelly 2018).

sions and back-office data," giving the CDO responsibility for "providing for the availability of common, usable, Defense-wide data sets." In response to this Congressional direction, in early 2020, the DoD Chief Information Officer (CIO), Dana Deasy, assumed responsibility for the CDO, and released a memo that "the CIO's office will take charge of all of the department's existing data governance bodies and create a new Data Governance Board" (Serbu 2020), a strong shift toward addressing the complexities introduced by having similar data referenced with different terminologies across DoD.

Simply put: businesses, institutions, and government agencies are transforming in response to a data-rich world; the transformations are ongoing and challenging; and barriers routinely include cultural or collective mindset.

However, even within this group mindset, we often face an additional barrier in the mindset of the individual. In the NewVantage survey foreword, Thomas H. Davenport and Randy Bean note that "[i]t is particularly striking that 77% of respondents say that 'business adoption' of big data and AI initiatives continues to represent a challenge for their organizations." Further, Davenport and Bean say, "Respondents clearly say that technology isn't the problem—people and (to a lesser extent) processes are. We hear little about initiatives devoted to changing human attitudes and behaviors around data. Unless the focus shifts to these types of activities, we are likely to see the same problem areas in the future that we've observed year after year in this survey" (NewVantage 2019).

Similar attitudes and individual mindsets are reflected among members of the defense acquisition workforce. As in industry, defense acquisition professionals use data in their everyday work. They, for example, monitor financial data, evaluate testing data, and create and process contracting data. Unfortunately, some people may not recognize their often significant roles and value within the data life cycle of a program as anything different from their usual tasks in the acquisition process, or inaccurately believe that data science is only the purview of data scientists with extensive technical backgrounds. The bottom line is that an acquisition professional does not have to be a data scientist to have a significant role in the data life cycle.

Their mindsets also tend to keep a static view of what data they use today and have always used instead of seeking new data that may help improve acquisition processes and insights. In addition, an acquisition professional who is not aware of his or her role in the data life cycle is unlikely to be aware of the other roles in the data life cycle—and may not recognize that other roles are not being performed, or are being performed inadequately.

As explained in Chapter 3, the data life cycle has 10 different phases: question, define, coordinate, generate, collect, curate and manage, analyze, visualize, disseminate and interpret, and assess. A gap or deficiency in any one of these phases can impact the quality of performance in all other phases. For this reason, even a lack of simple awareness of the data life cycle has the potential to undermine the quality of decision making in the acquisition process.

> **Finding 5.1:** An ever-increasing majority of the acquisition workforce is participating in the data life cycle. Just as defense acquisition is most successful when it is accomplished collaboratively, so is data science. Data science is an inherently collaborative field where success requires cooperation, communication, and coordination. Phases of the data life cycle directly or indirectly depend on one another, and those participating in the data life cycle must be able to relay key details to others. (See Box 5.1.)

Having a sense of how data matures at each step of its life cycle and who may or may not have responsibility for the data at any given point in its life cycle will likely be a new mindset for the broader acquisition community to understand and appreciate. Without an understanding of how they fit in a broader data life cycle and how their actions affect other parts in the cycle, the defense acquisition workforce may be missing or slowing opportunities to improve acquisition as well as advance their careers.

Connected to these challenges in institutional and functional culture—and lurking in the background—is a perceived distinction between those that have data science skills and those that do not. "I am not a math person," is a common refrain and is indicative of a substantial barrier in STEM education. As was noted in 2016, "The culture of science, technology, engineering, and mathematics (STEM) education has an effect on many students' interest, self-concept, sense of connectedness, and persistence in

---

BOX 5.1
Critical Moments for Communication in the Data Life Cycle

- Question(s): Questions must be understood among all who are facilitating the data life cycle for a project. Terms need to be defined and objectives must be clear.
- Collect: How the data were collected and curated dictates what kind of analyses can and should be done.
- Uncertainty: Limitations and uncertainty found in phases of the data life cycle must be communicated during interpretation, dissemination, and prior to decision-making.

these disciplines" and "STEM 'gateway' courses continue to negatively impact STEM student persistence" (NASEM 2016 p. 59). Thus, members of the current and future DoD acquisition workforce may exhibit discomfort with data science due to cultural, institutional, and educational barriers. Many may also not be aware of how they already interact with and use data regularly to make data-informed decisions. And, unfortunately, despite having an aptitude and/or initial interest, it is important to recognize that DoD may need to help individuals overcome data-centric deficiencies in their education and other circumstances beyond their control that may also be affecting their mindset toward data and its use. These barriers especially affect those who identify as a woman or an ethnic or racial minority (NASEM 2016).

Conclusion 5.1: Defense acquisition workforce members' awareness of their value and individual role in the data life cycle is critical for data science applications. Barriers to this awareness include (1) lack of familiarity and use of the data life cycle, (2) adoption challenges for data science technologies and practices, and (3) widespread self-doubt of their own skillsets and insecurity or anxiety with STEM (generally) and data science (in particular).

SKILLSET

As DoD contemplates preparing its workforce to more fully embrace and incorporate data science, it is worth considering the types of skills needed to develop and shape decision-informing data insights. As shown in Appendix B, there is no acquisition career field named "data scientist" (DoD 2019). There is also no federal career field identified as a data scientist, although in 2019 the Office of Personnel Management (OPM) did allow agencies to add data science titles to a number of positions within their organizations (Wagner 2019). Essentially, the data science tag could apply to jobs within existing federal job types, including operations research, statisticians, cost analysts, or IT specialists. This "job title" approval essentially set the stage and expectation that data scientists do not work for themselves or operate in isolation looking for "good ideas," but are part of a team and will be embedded with a "domain" of some kind needing or wanting to make better data-informed decisions for the questions they have.

With this permission to establish "unofficial" data scientist job titles, OPM Associate Director for Employee Services Mark Reinhold wrote that "[d]ata scientist work is multifaceted and requires talent from interdisciplinary backgrounds." He went on to say that "[d]ata scientists are defined as practitioners with sufficient knowledge in the areas of business needs, domain knowledge, analytical skills and software and systems engineering to manage the end-to-end data processes in the data life cycle" (Wagner 2019). That said, there is a distinction between the "data scientist" role and the range of skills needed throughout the data life cycle that could be provided by other functions in the acquisition community.

Before delving into the broader data life cycle skillset, the committee highlights two critical considerations from Chapter 3.

- The data life cycle is a bi-directional workflow or process.
- For any given project, full utilization of all the phases of the data life cycle requires varied skills, and a single person is unlikely to have all of those skills.

From these, the committee concludes the following:

Conclusion 5.2: Improved data use in acquisition requires a collective skillset and teamwork among the acquisition workforce members. Team members will need data skills appropriate for their functional role in the acquisition community.

Data Literacy

In response to the growing importance and ubiquity of data and data-related tasks, the private sector is increasingly prioritizing *data literacy* for all employees, not just those in data analytics or data science roles. When it comes to the U.S. workforce, "data isn't used in a vacuum: it touches many other roles, and those employees need the literacy to handle it effectively" according to Laurence Bradford, creator of Learn to Code with Me. Amy O'Connor, chief data and information officer at Cloudera conveyed to Bradford in a 2018 interview that "organizations need a broad set of data skills, and they need to be in various different roles across the organization." CDO roles, such as O'Connor's, are now common in the workplace, and their responsibilities have grown over time. They often report that "poor data literacy" is among the top challenges in achieving company goals (Bradford 2018).

Similarly, higher education is focusing on defining data literacy and developing related training since data-centric skills contribute to almost every discipline. Some institutions of higher education are, in general, shifting to systems wherein students demonstrate core competencies (as opposed to completing specific courses). Data and information literacy tend to appear on these short lists of core competencies along with more common competencies such as reading/writing literacy and quantitative literacy.

Using slightly different language, the National Academies 2018 report on *Data Science for Undergraduates: Opportunities and Options* recommended that some form of data science be taught to all undergraduate students:

> *To prepare their graduates for this new data-driven era, academic institutions should encourage the development of a basic understanding of data science in all undergraduates. (Rec 2.3)*

Exposure to the data life cycle now occurs in introductory general education courses, undergraduate and master's degrees in data science and related topics, and online courses and certificates for a rapidly growing number of students in two-year and four-year colleges (NASEM 2018).

Finding 5.2: Upon graduation, undergraduate students are increasingly expected to have learned and applied some level of data science knowledge and skills.

Conclusion 5.3: Defense acquisition can expect the future workforce to have a baseline set of data science skills.

What these skills are is more difficult to define. Because we are in the early stages of curriculum and program development for data science, it is difficult to build consensus. Borrowing from NASEM (2018), here the committee refers to data literacy as a collection of baseline data science skills similar to those that would be taught at the undergraduate level (Box 5.2).

Data literacy can be achieved in a variety of ways; ideally, there are multiple pathways to data literacy that accommodate all learners that include concepts common to many introductory data science and data literacy courses. For example, during the April 2020 workshop on "Improving Defense Acquisition Workforce Capabilities in Data Use," Matthew Rattigan (University of Massachusetts Amherst) stated that data literacy includes exploratory data analysis skills, data visualization and summarization skills, familiarity with experimental design, conceptual comparison of prediction vs. causality, the scientific method, and an understanding of data ethics, fairness, transparency. While differing slightly from the National Academies definition in Box 5.2, there remains a strong emphasis on how data are collected, stored, and used as well how to communicate with data.

Finding 5.3: While specific outcomes are being updated and refined, *data literacy* for many undergraduate students includes an understanding of the data life cycle and how it works, data story-telling and communication, and an ability to recognize matters of ethics, privacy, and security.

> **BOX 5.2**
> **Skills and Concepts Needed to Achieve Data Literacy**
>
> Data literacy concepts include data questions; assessment of data sources and quality; margin of error and uncertainty; concerns related to ethics, security, and privacy; basic tools and techniques for visualizing and analyzing data; interpreting visualizations and results; and best practices for communication.

As data literacy continues to proliferate and relevant training becomes more widely available in higher education, or becomes available within the DoD education and training community, the acquisition workforce should be given similar opportunities.

Recommendation 5.1: The Department of Defense and its components should ensure that all members of the acquisition workforce and its leadership eventually have a common baseline data literacy, which includes an understanding of the data life cycle and how it works, data story-telling and communication, and an ability to address matters of data ethics, privacy, and security—all skills that may evolve along with the use of data science in government, industry, and academia.

Six Roles for the Data Life Cycle

Workforce-wide data literacy is necessary but not sufficient for improved data use within the defense acquisition workforce. To see why, it is important to understand the data, analyses, and—most importantly—decisions that are made in acquisition as discussed in Chapter 4. For example, Anton et al. (2019a) argue that teams are required to make these decisions and members of the team have different levels of technical skills. In January 2020 testimony to the National Academies, Sallie Ann Keller noted that "data science is a team sport" and "there are many levels of data acumen." Further discussion of team structures and roles is in the section below titled "Team Structures for Data." Across industry and government, and as taught in academia, executing the data life cycle generally requires multiple roles with varying skillsets. Keller and Bethany Blakey alerted the committee to notable data science roles at the National Institutes of Health in their testimony in January and March 2020.

The defense acquisition workforce may not find the following six job titles in its workforce, but the six roles that follow describe the data-related skillsets required to support a project utilizing the data life cycle.

Data Engineers

Data engineers are specialists with a technically specialized skillset. Typically, there are few data engineers within an organization. They tend to be matrixed across teams or deployed to support critical domain teams. Data engineers are broadly responsible for supporting the best methods, tools, and interfaces to prepare data and make it accessible. They establish and curate data sets so that they are maintained and available for subsequent analysis and decision making. They create the data, computing, networking, memory, and storage platforms (often called the "data enterprise"). They support cloud and on-premises storage and computing, manage data security, develop systems specifications, implement system interfaces to access, retrieve, and process data, and support the data architecture implementation using the big data tools identified for the data analytics platform that allows to access, integrate, and store data.

Data Scientists

Data scientists are also specialists. They sometimes lead a data science team and have a technically specialized skillset and deep analytics knowledge. Their skillset spans the data life cycle, with emphasis on advanced techniques for data collection, curation, management, analysis, and visualization. Like data engineers, organizations employ few data scientists, and they tend to be matrixed across teams or deployed to support critical domain teams. Because debate continues regarding skills needed to be identified as a data scientist, data science degrees from different institutions can equip students with very different (but useful) skills. In general, students from data science programs will learn mathematical, computational, and statistical foundations; data management and curation; data description and visualization; data modeling and assessment; workflow and reproducibility; communication and teamwork; domain-specific considerations; and ethical problem solving. Their academic experience will typically require exposure to real-world problems, data, tools, and ethical considerations with different programs sometimes aligning with a specific domain expertise.

The committee emphasizes that the role of data scientists will likely not be ubiquitous within the acquisition workforce. Expectations for skills for data scientists have evolved significantly over the past decade. Early on, as is noted in Chapter 3, there were discussions about what skillsets defined data science and whether it was a unique field of study or a natural evolution of data-intensive domains such as statistics or computer science. With these discussions came a reckoning (and a slow acceptance) that the expectations for data scientists were different and often more interdisciplinary than what members of the U.S. workforce saw in traditional academic programs. Appendix D details the skillset needed to achieve mastery of *data acumen* presented in the National Academies 2018 report *Data Science for Undergraduates: Opportunities and Options*.

Finding 5.4: Data scientists are experts across the data life cycle, with special emphasis on advanced techniques for collection, curation, management, analysis, and visualization.

Finding 5.5: Due to an evolving data science curriculum in higher education, not all data science degrees prepare students with the same skillsets.

Data Analysts

Across the acquisition workforce, staff use data to conduct analyses in support of acquisition functions. These analysts are commonly embedded in domain teams. In acquisition, a domain team might be an aircraft cost estimating team, or a contractor-specific quality assessment team, or a missile test and evaluation team, for example. Data analysts on these teams have skills in the analysis, visualization, interpretation, and communication of data. They obtain, clean, and transform data in preparation for specific analysis. Analysts compute summary and descriptive statistics (e.g., measures of frequency, central tendency, variation, and position), perform statistical modeling (including contingency tables and regressions), and apply more advanced statistical and machine learning methods to calculate estimates and uncertainties. They create both static and time-dependent visualizations. They implement sampling strategies, design surveys and statistically efficient experiments, and communicate analytic results in a specific domain. Within DoD, data analysts are often aligned with specific acquisition functions such as cost estimating and pricing, test and evaluation, and logistics.

Data Users

Staff across the acquisition workforce utilize data in their day-to-day work, whether they are generating, collecting, or interpreting data to create value for the organization. In the context of their specific roles, data users require a baseline set of data science skills, that is, data literacy, including familiarity with basic terminology, understanding data limitations and their effect on decisions, and the ability to read charts and graphs. Acquisition program managers, contracting officers, and negotiators are just a few of the many acquisition professionals who use data on a regular basis.

Domain Experts

Domain experts have mission-related subject-matter expertise, with skills contextual to their area of interest. Domain experts are key team members who work with data scientists, engineers, and analysts to tailor collection, management, and analyses to support key decisions. Skills also include interpreting, disseminating, and communicating results from the data life cycle. Domain experts understand the context of the data in their domain, where it comes from, what it means, and how it is used. For example, a contract or program expert will understand schedule or contractor cost performance data associated with their program. A logistician may understand data associated with supply chains or distribution channels. A test and evaluation expert will know how to instrument a missile in order to collect telemetry during missile performance.

Leaders and Decision Makers

Leaders and decision makers have or need data-related skills that fall into three broad categories: building a culture that values data; governing, managing, and protecting data; and promoting efficient and appropriate data use. Additionally, leaders and decision makers need skills in data visualization, interpretation, and communication. Leadership skills also include identifying data needs, championing data use and shaping institutional cultures to embrace data, prioritizing and funding data governance, recognizing the value of data assets, aligning quality with intended use, increasing the capacity for data management and analysis, and understanding limitations and uncertainty. The section below titled "Team Structures for Data Science" addresses the skills for supervision and management of data science projects and teams.

> **Finding 5.6:** Executing the data life cycle is a collaborative endeavor and generally requires a collective skillset found in teams of data engineers, data scientists, data analysts, data users, domain experts, and leaders/decision makers.

TOOLSET

For the DoD acquisition system to achieve the benefits of an empowered workforce in data use, it must have access to the full range of data science capabilities—from data generation, collection, and curation to data analysis, visualization, and dissemination.

DoD has capabilities in data analysis and analytic tools, especially through analytic organizations such as the Office of Cost Assessment and

Performance Evaluation (CAPE); the Center for Army Analysis, the Office of the Chief of Naval Operations Assessment Division (N81); the Air Force Chief Analytics Officer (A9); the Office of Acquisition, Analytics and Policy in the Office of the Under Secretary for Acquisition and Sustainment; the Office of People Analytics in the Office of the Under Secretary for Personnel and Readiness; and the Analytics Center of Excellence in the Defense Logistics Agency. Analysts are embedded with leaders in military and civilian organizations and entities throughout DoD, including acquisition commands and program offices. In addition, DoD has ready access to additional analytic support through outside organizations, including FFRDCs, University Affiliated Research Centers (UARCs), and contractors in the defense industrial base and the academic research community.

While DoD has had a chief information officer for decades, DoD has not had enterprise-level leaders dedicated solely to the collection, storage, or curation of data. In the past few years, as is noted earlier, DoD has established new positions for CDOs, with responsibility for managing DoD data assets, including the standardization of data format, the sharing of data assets, and the development of common, usable, defense-wide data sets. However, the acquisition data needs of a department in which more than 150,000 designated acquisition personnel (along with others in support of acquisition) in hundreds of organizations spending more than $300 billion annually[2] are unlikely to be addressed by a single official who is largely limited to trying to persuade others to invest in data resources and data sharing. Significant investment—on the order of billions of dollars a year that DoD currently spends trying to achieve an auditable financial statement—will likely be needed to develop the structured databases and modern analytic tools needed to build a modern data-centric environment inside DoD. Estimates of total spending on modern data-centric environments or data systems are difficult to obtain based on how budgets are tracked. However, Anton et al. (2019b) estimated that DoD budgeted about $0.5 billion for major acquisition IT systems and about $2.5 billion for logistics and supply-chain management systems in FY 2017 based on public Select and Native Programming—Information Technology (SNaP-IT) budget-request exhibits. This constitutes a reduction from the $4 billion or so budgeted for these systems going back to FY 2008 (Anton et al. 2019b). DoD recently established a new ADVANA platform for assembling critical data sets and making them available for analysis, but far more investment

---

[2] The enacted DoD budget in FY 2020 included $248 billion for Research, Development, Test, and Evaluation (RDT&E) and Procurement (Office of the Under Secretary of Defense (Comptroller)/Chief Financial Officer 2020). In addition, acquisition includes spending a significant portion of the $290 billion budgeted for Operation and Maintenance and $17 billion for Military Construction for FY 2020.

is still needed to bring comprehensive data sets into the system and to make them available for data analysis.

Conclusion 5.4: A data-capable defense acquisition workforce must—as a whole—have access to the full range of data science capabilities, from data collection and curation to data analysis and visualization.

TEAM STRUCTURES FOR DATA SCIENCE

From the section "Six Roles for the Data Life Cycle," we know that executing the data life cycle is a collaborative effort. Indeed, a data science project generally requires teamwork and coordination among the team's six data roles. Within a single organization, teams will differ—often dramatically. So, what constitutes a team for a data science project? How are they structured? How are they led?

When leveraging data as a strategic asset, it might be tempting to think that everyone needs to be a data scientist or data engineer. These technical roles are indeed critical for executing the data life cycle. Data scientists and data engineers require advanced training, and there is and will continue to be a shortage of data engineers and data scientists (Davenport and Patil 2012) in the U.S. workforce broadly. Accordingly, organizations struggle to find or hire data scientists and data engineers within their workforces. While it may not be many, the defense acquisition system, like every organization, will need data scientists and data engineers, and it will encounter challenges to employing them. Options for increasing the number of data engineers and data scientists within an organization include upskilling, hiring, and contracting. Outside contractors can provide added capabilities, but data restrictions and inherently governmental functions could limit their use.

While all six roles are critical for defense acquisition communities for executing the data life cycle, the vast majority of the data-centric workforce will not need complex technical skills. *Typically, organizations of all types and sizes should have many data analysts, data users, and domain experts and fewer data engineers, data scientists, and leaders/decision makers.* For DoD, the vast majority of the defense acquisition workforce can be identified as data analysts, data users, or domain experts, though identifications may change project to project, or with additional training.

In Table 5.1, the committee maps the data life cycle from Chapter 3 to the six roles and their skills from this chapter. Here, the committee notes that data scientists often oversee and support the entire process. The committee also notes that two or more roles overlap at every phase, and data users, domain experts, and leaders/decision makers are involved in similar phases albeit at the level of data literacy versus a deeper acumen mastery level.

TABLE 5.1 Workforce Roles within the Data Life Cycle

Role	Phases in the Data Life Cycle										Data Science Capability Level
	Question	Define	Coordinate	Generate	Collect	Curate and Manage	Analyze	Visualize	Disseminate and Interpret	Assess	
Data engineer		implementation specialist	implementation specialist	implementation specialist	implementation specialist	implementation specialist					
Data scientist	assess/refine	science and theory	science and theory	science and theory	science and theory	science and theory	deeper capabilities	deeper capabilities	deeper capabilities	deeper capabilities	acumen
Data analyst	assess/refine						deeper capabilities	deeper capabilities	deeper capabilities	deeper capabilities	
Data users	assess/refine							deeper capabilities	deeper capabilities	deeper capabilities	
Domain experts	pose										
Leaders, decision makers, and managers	pose										literacy

Leaders and decision makers—including those with non-technical backgrounds or training—can effectively manage data science projects by using common strategies and approaches for managing collaborative, cross-functional projects that include technical aspects. However, supervising and managing data science projects requires:

- Valuing data and analysis in decision making;
- Valuing and understanding the power and limitations of data use and analysis;
- A basic familiarity with the data life cycle;
- Ability to argue, make a case, or tell a story using data or analytic results (visualization);
- An ability to understand and interpret data and analytic results, and communicate them to various audiences; and
- An aptitude for asking the right questions of the team—both to determine which data and analytics are viable and to inform a decision.

Leaders and decision makers should routinely ask questions such as the following: What are the key questions that need to be answered? What data do we need? What data do we have available to us? How can we access these data and what are the related challenges? Are there data ethics, privacy, and security concerns, and if so, how might we address them? What is the quality of our collected data? What are the data's limitations and opportunities? What are the data telling us, and how do you know? Are there limitations to inferences that we make from this project? What is the uncertainty involved in this analysis? If the questions are not answered, what additional data are needed to answer those questions?

Leaders must identify the critical data skills needed for their organization and each project by assessing current staff capacity, performing a data skills gap analysis, identifying ways to meet those needs, and making investments to right-size the team.

Conclusions 5.5: Management of data science projects uses strategies and approaches for leading collaborative, cross-functional, technical projects with specific attention paid to the development of a team that has skills across the data life cycle and to asking questions specific to the quality and utility of data.

There are a variety of structures for teams executing the data life cycle, each intended to foster collaboration and knowledge creation and dissemination through the organization. As is outlined in Box 5.3, the type of optimal team structure will depend on the structure and goals of the enterprise or organization.

> **BOX 5.3**
> **Choosing Team Structures for Executing the Data Life Cycle**
>
> A centralized model should be considered when:
> - Data are enterprise-wide;
> - It is difficult to recruit and retain data scientists and data engineers; and
> - Data use opportunities are generally homogeneous.
>
> An embedded model should be considered when:
> - Data are specialized;
> - Sustained domain expertise and experience are valued; and
> - Data use opportunities are generally dynamic.

In a *centralized model*, data scientists and data engineers are grouped together and often serve at least one of the two following functions: (1) an innovation or research hub that supports the development of new prototypes or processes, and (2) an applied support center assigned to different functions for a fixed period of time or project. In these roles, data scientists and data engineers may need to adjust to several domains and business contexts and to emphasize communication skills. However, being centralized with other data scientists and data engineers can better support leveraging data for enterprise-wide innovation.

In an *embedded model*, each division or group within an organization hires its own data scientist(s). Here, data scientists generally have domain knowledge acquired over time spent in the division, increasing their effectiveness and quality of support. However, tools, resources, and practices tend to be heterogeneous across the entire organization, which can be more difficult to innovate data practices or optimize the use of data across an organization with siloed data scientists in limited communication. Hybrid models, of course, exist and organizations customize models to meet their needs.

However DoD or individual military departments or acquisition programs choose to organize its data science teams, the acquisition workforce should have a collective data life cycle mindset, skillset, and toolset. Each member of the workforce will need well-defined roles within the data life cycle for any given project, and a set of accompanying skills for that role. Teams will need to be established, customized, nurtured, and managed such that data use is embedded in all acquisition processes. Leaders will need familiarity with the data life cycle, management skills that optimize their data science talent, and a commitment to data-informed decision making. Data science has and will continue to evolve at a rapid pace, and so too will best practices and approaches for supporting a workforce empowered to use data. These recommended workforce characteristics and structures reflect current best practices and approaches in industry, government, and academia.

Recommendation 5.2: The Department of Defense should prioritize the utilization of data and the data life cycle by appropriate and judicious investment in the data science mindset, skillset, and toolset of the acquisition workforce.

This chapter concludes with Box 5.4, which revisits the defense acquisition examples first introduced in Chapter 1, with additional focus on the data-related skills and roles necessary in each situation that contributed to its success.

**BOX 5.4
The Power of Data:
Examples in Defense Acquisition**

Enabling Multi-Year Appropriation Approval and Program Benchmarking

The Cost Assessment Data Enterprise (CADE) system (operated by the Office of the Secretary of Defense's [OSD's] Cost Assessment and Program Evaluation [CAPE] directorate) was enabled through data-valued leadership of the CAPE Director, program-related data generation by defense industry members, coordination with industry on data infrastructure and storage, a CAPE-built data collection tool. Previously, pdf documents were manually entered into a database used exclusively by CAPE analysts. For CADE, the CAPE analysts changed the prior practice of using pdf documents and designed a system around machine-readable formats. The related curation and management were done by contracted data engineers. Initially CAPE analysts used the data for cost-estimates and multi-year contract decisions. However, over time, the data were also made available for other analyses in OSD, service programs, and staff offices. As a result, cost estimation across the entire DoD became more aligned and defensible.

Saving Billions of Dollars with Category Management

The Air Force Category Management (CM) was envisioned by Air Force contracting leaders in the early 2000s, and is now embraced at Air Force headquarters under the Deputy Under Secretary of the Air Force, Management and Deputy Chief Management Officer, Office of the Under Secretary of the Air Force. CM is not a contracting process. It informs early acquisition decisions and may result in centralized or decentralized acquisition strategies, completely change the Air Force's buying approach for a product or service, or may deliberately decide to make no changes to the way a requirement is met. Categories of services or supplies are defined by requirements owners (domain experts), such as civil

BOX 5.4 Continued

engineers, security forces, the medical community, or information technology. Historical spending needs data for these requirements are extracted from existing government data systems (FPDS-NG) by procurement specialists and analysts and domain experts; data engineers were hired to apply more sophisticated visualization techniques enabling analysts to characterize timing of and magnitude of historical buys to inform a CM decision. As the data are examined, an acquisition strategy will take shape and sometimes broader strategic opportunities may be exposed. For example, the security forces team coordinated its need for working dogs with other federal agencies resulting in a central acquisition for multiple federal agencies. Leadership attention, dedicated domain expertise, and funding of specialized talent such as data engineers operating in a centralized organizational structure have made Air Force CM successful.

Improving Logistics and Operational Availability of Military Equipment

The U.S. Army's Velocity Management initiative improves the analysis, dissemination, and interpretation phases of the data life cycle for availability and maintenance data for military equipment. Because no single "process owner" had authority to reform the whole logistics system, a coalition of stakeholders, logistics providers, technical experts, line managers, and FFRDC (federally funded research and development centers) researchers worked together to assess the logistics ecosystem, design new logistics processes, measure success, and iteratively reassess and improve the logistics system (Dumond and Eden 2005). The external FFRDC domain and analyst experts were an important element to help support out-of-the-box thinking, but so were the existing logistics domain experts, data analysts, leaders, and decision makers. Metrics relied on existing data sources, but they were "combined and used in innovative ways" to enable Velocity Management (Dumond and Eden 2005, p. 224). Additional information had to be collected (including financial and information flows) to understand how the process worked and what steps were involved in the existing system.

Saving Lives with Test and Analysis Programs

Operational test and evaluation are inherently data-driven processes. Accordingly, every phase of the data life cycle is used in this community. Impacts of operational test and evaluation, however, are particularly striking because they are directly responsive to critical questions, for example, on the safety of helmets. In this way, the question phase of the data life cycle is paramount. Test and evaluation also rely heavily on skills from data engineers, to design data collection and prepare data for analysis, and data scientists, to perform statistical design experiments and formal inference and uncertainty quantification to address requirements. Both embedded and centralized models are successfully employed to provide specialist expertise. Leading test and evaluation projects does not necessarily require deep technical expertise, but it does require developing and leading a collaborative team with skills across the data life cycle.

continued

> **BOX 5.4 Continued**
>
> **References**
>
> DOT&E (Director Operational Test and Evaluation), 2016, "Director, Operational Test and Evaluation (DOT&E) FY 2016 Annual Report, https://www.dote.osd.mil/Publications/Annual-Reports/2016-Annual-Report/.
>
> Dumond, J., and R. Eden, 2005, "Improving Government Processes: From Velocity Management to Presidential. Appointments," RAND Corporation, https://www.rand.org/pubs/reprints/RP1153.html.
>
> Westermeyer, R., 2017, "Bullet Background Paper on Air Force Installation Contracting Agency's Business Analytics Capability," Wright-Patterson Air Force Base, Ohio.
>
> Westermeyer, R., 2019, "Bullet Background Paper on Business Reform—Category Management." Wright-Patterson Air Force Base, Ohio.

REFERENCES

Anton, P.S., M. McKernan, K. Munson, J.G. Kallimani, A. Levedahl, I. Blickstein, J.A. Drezner, and S. Newberry. 2019a. *Assessing Department of Defense Use of Data Analytics and Enabling Data Management to Improve Acquisition Outcomes*, Santa Monica, CA: RAND Corporation, RR-3136-OSD, August. https://www.rand.org/pubs/research_reports/RR3136.html.

Anton, P.S., T. Conley, I. Blickstein, A. Lewis, W. Shelton, and S. Harting. 2019b. *Baselining Defense Acquisition*, Santa Monica, CA: RAND Corporation, RR-2814-OSD. https://www.rand.org/pubs/research_reports/RR2814.html.

Bradford, L. 2018. "Why All Employees Need Data Skills in 2019 (and Beyond)." Forbes. October. https://www.forbes.com/sites/laurencebradford/2018/10/11/why-all-employees-need-data-skills-in-2019-and-beyond/?sh=6329b8c6510f.

Conlin, M. 2020. Presentation at the Workshop on Improving Defense Acquisition Workforce Capability. Virtual. April 14.

Davenport, T.H. and D.J. Patil. 2012. "Data Scientist: The Sexiest Job of the 21st Century. Harvard Business Review. https://hbr.org/2012/10/data-scientist-the-sexiest-job-of-the-21st-century.

"Defense Acquisition Workforce Key Information, Overall," briefing, Washington, D.C., FY20Q3 (30 June 2019). https://www.hci.mil/docs/Workforce_Metrics/FY20Q3/FY20(Q3)OVERALLDefenseAcquisitionWorkforce(DAW)InformationSummary.pptx.

Hadziristic, T. 2017. "The State of Digital Literacy in Canada: A Literature Review." Brookfield Institute for Innovation + Entrepreneurship. April.

Huynh, A., and A. Do. 2017. "Digital Literacy in a Digital Age." Brookfield Institute for Innovation + Entrepreneurship. August. https://brookfieldinstitute.ca/wp-content/uploads/BrookfieldInstitute_DigitalLiteracy_DigitalAge-1.pdf.

Kelly, W. 2018. "Being Digitally Savvy in a Digital World." Rural Development Institute. February 26. https://medium.com/@rdi_77976/being-digitally-savvy-in-a-digital-world-b7bb291be85f.

MediaSmarts. 2016. "Digital Literacy Fundamentals." http://mediasmarts.ca/digital-media-literacy-fundamentals/digital-literacy-fundamentals.

NASEM (National Academies of Sciences, Engineering, and Medicine). 2018. *Data Science for Undergraduates: Opportunities and Options*. Washington, DC: The National Academies Press. https://doi.org/10.17226/25104.

NASEM. 2016. *Barriers and Opportunities for 2-Year and 4-Year STEM Degrees: Systemic Change to Support Students' Diverse Pathways*. Washington, DC: The National Academies Press. https://doi.org/10.17226/21739.

NewVantage Partners. 2019. "Big Data and AI Executive Survey 2019: Executive Summary of Findings." NewVantage Partners. http://newvantage.com/wp-content/uploads/2018/12/Big-Data-Executive-Survey-2019-Findings.pdf.

Orad, A. 2020. "Why Every Company Is a Data Company." Forbes. https://www.forbes.com/sites/forbestechcouncil/2020/02/14/why-every-company-is-a-data-company/?sh=6ce301ef17a4.

Serbu, J. 2020. "Pentagon Racing to Establish New Chief Data Officer Within CIO's Office." Federal News Network. January 28. https://federalnewsnetwork.com/defense-main/2020/01/pentagon-racing-to-establish-new-chief-data-officer-within-cios-office/.

Van Deursen, A.J.A.M., E.J. Helsper, and R. Eynon. 2014. M*easuring Digital Skills. From Digital Skills to Tangible Outcomes Project Report*. https://www.oii.ox.ac.uk/research/projects/?id=112.

Wagner, E. 2019. "OPM Announces New 'Data Scientist' Job Title." Government Executive. July 1. https://www.govexec.com/management/2019/07/opm-announces-new-data-scientist-job-title/158139/.

Yueh, J., and R. Bean. 2018. "Every Company is a Data Company." Forbes. https://www.forbes.com/sites/ciocentral/2018/09/26/every-company-is-a-data-company/#25f21bcd5cfc.

6

Preparing and Sustaining a Data-Capable Defense Acquisition Workforce

Workforce development in data analytics and data science is critical to the Department of Defense's (DoD's) current and future acquisition functions, and it supports a broader DoD effort to make the department a data-centric organization, as outlined in the 2020 DoD Data Strategy (DoD 2020). While not all defense acquisition workforce members will or should be data scientists, becoming a data-centric organization depends on personnel that understand the value of data, the conditions that affect the quality of the data, and how data can better inform decisions in the acquisition environment. The key to success, however, will be attracting, developing, managing, and retaining personnel with robust data skills and experience; whether that be as a military member, civil servant, or a contracted resource.

A review of relevant training opportunities currently available to personnel provides a crucial background for the development and adoption of future programs and strategies. The committee undertook site visits to Defense Acquisition University (DAU), the Air Force Institute of Technology (AFIT), and the Naval Postgraduate School (NPS) to gather key information on current and prospective data training efforts available to acquisition personnel and leaders (see Appendix A). The committee also solicited input on data upskilling from organizations outside DoD through site visits and invited testimony. Taken together, information gathered from the site visits and invited speakers provided key insights for future data analytics and data science training programs for acquisition personnel.

OVERVIEW OF TRAINING APPROACHES IN DATA SCIENCE AND ANALYTICS

Given the rising use of data across many sectors and domains, it is important to recognize the variety of training approaches currently available. Many of these programs are evolving as new technologies and data platforms become available. Demand for data talent is expanding across industries, driving demand for multiple models of training delivery. Importantly, without operational context and an understanding of the specific programmatic needs, one type of training is not better than another, and therefore the committee does not endorse any specific training program or approach. However, understanding the characteristics of various training approaches can provide opportunities for acquisition professionals and managers to map training programs to their data needs and available resources. Below are brief descriptions of training approaches currently available for increasing data capabilities. Table 6.1 provides a summary of these training options and thus can be a guide for determining which platforms or approaches are appropriate for meeting specific data needs.

Higher Education

Higher education institutions, including 2-year and 4-year colleges, offer a variety of data-science training opportunities. They include traditional courses and degree offerings spanning associate's, bachelor's, master's, and doctoral programs. Higher education institutions are externally and independently accredited to ensure that recognized standards are met. For most degree programs, learners must apply and be accepted into the program, but individual courses offered through continuing education programs may only require prerequisite courses for admittance. The majority of the data analytics and data science courses currently offered in higher education institutions occur over the span of a semester or quarterly schedule and are either in person, online, or in a hybrid model that synchronously combines in-person and virtual attendance. Currently, there are efforts for developing ABET (Accreditation Board of Engineering and Technology) accreditation of data science programs. Course fees range depending on the institution, but they can be costly if not offset by scholarships or employer contributions. Importantly, the data courses and programs in non-defense higher education institutions are not tailored for defense acquisition applications and scenarios.

At the time of this report's publication, higher education is undergoing substantial adaptation as a result of the COVID-19 pandemic. Given safety precautions associated with in-person learning during the pandemic, higher education has rapidly adapted to increase access to courses through delivery

TABLE 6.1 Characteristics of Different Training Offerings for Data Analytics and Data Science

	Cost	Asynchronous	Part-time	Human graded	Human supported (e.g., teaching assistants)	Program/Degree	Content delivery (Virtual, Hybrid, or In Person)	Accreditation (External, Internal, or Non)	Certifying exam or assessment	Organization-level personalization
Traditional Higher-Education: Degrees	high	varies	varies	typically	yes	yes	in person, hybrid, and virtual	externally accredited	at course level	rarely
Higher-Education: Continuing Education and Auditing	varies	rarely	yes	varies	yes	varies	typically in person	externally accredited	at course level	rarely
Certificate Programs	varies	varies	varies	typically	yes	varies	typically in person	varies	yes	sometimes
Micro-credentials	varies	varies	yes	varies	yes	varies	both	internally	yes, varies	rarely
Bootcamps	high	rarely	varies	varies	yes	not really	in person	internal	varies	rarely
Massive Open Online Courses (MOOCs)	free to low	typically	yes	rarely	rarely	not by design	virtual	rarely	if paid for, varies	rarely
Executive/Leadership Programs	varies	varies	yes	N/A	yes	no	typically in person	rarely	rarely	yes

modes that allow for fully remote and hybrid participation.[1] Institutions have expanded digital infrastructure to support multiple education delivery modes, and are adapting pedagogical approaches that support digital learning environments. The adaptations not only protect the health of faculty and learners, but also connect institutions to a broader range of learners. The educational adaptations developed in response to the pandemic are creating a window of opportunities for higher education institutions to collaborate with a broader set of external stakeholders and organizations. Higher education is primed to partner with defense acquisition to provide accessible and tailored course offerings in data analytics and data science.

Certificate Programs

Over the past decade, there has been substantial growth in certificate programs in data science and data analytics. The growth in data certificates has occurred in both the higher education and commercial sectors, and some organizations have developed their own in-house certification programs. Typically, certificate programs are a collection of courses that can be delivered either in person or online and take less time to complete than traditional academic programs. However, some certificate programs are self-paced, and completion of a certificate can span from a few weeks to two years. Admission into a data science certificate program operated by academic institutions is similar to applying for an undergraduate program. Fees for certificate programs range across institutions, but they often require less investment than a graduate degree program.

Certificate programs are usually geared toward professionals that already have some business or computer science experience and can serve as an indicator of proficiency in data science concepts. They can also offer opportunities for learners to broaden their data analysis techniques and master new data tools. It is important to note that data certificate courses are generally not easier than master's degree courses. In fact, learners in higher education data science certificate programs may be enrolled in graduate-level courses, simply taking fewer courses than required for a graduate degree.

Data Bootcamps

Data analytics and data science bootcamps are typically characterized as short, intensive programs designed to quickly teach a combination of

---

[1] "Here's Our List of Colleges' Reopening Models" from *The Chronicle of Higher Education* available at https://www.chronicle.com/article/heres-a-list-of-colleges-plans-for-reopening-in-the-fall/ (as of March 11, 2021).

theoretical concepts and applications through interactions with the full data life cycle. Content can be delivered either in person or virtually in data bootcamps and typically involve 6–15 weeks of instruction. Some bootcamps are project-based, meaning the learners complete a project that demonstrates their abilities, as an immersive experience where learners can refine and apply their skills.

While some higher education institutions offer bootcamps for data analytics and data sciences, most are offered in the commercial sector at variable cost ranging from $25 a month to nearly $20,000 for a full program. A strong appeal of some data bootcamps are promises of job interviews or placements after successful completion of the program. However, bootcamps are not accredited, and it may be difficult to ensure the programs meet the recognized standards associated with higher education institutions.

Micro-Credentials

Micro-credentials are a form of specialized certification that verifies or attests to the proficiency of specific skills or competencies. They differ from traditional degrees and certificates in that they are generally offered in shorter or more flexible time spans and are more narrowly focused. Micro-credentials can be offered online, in the classroom, or in hybrid approaches through higher education institutions and third-party providers. Sometimes called digital badges or nano-degrees, micro-credentials can also be "stacked" by building upon one another as the learner continues mastering specific skills or competencies.

There are many benefits to using micro-credentials for increasing data analytic skills, including their specificity, short duration, flexibility, affordability, and personalization. Micro-credentials can also serve as an indicator of proficiencies and professional development if they are recognized by the learner's industry or employer. Depending on the provider, micro-credential courses are free to low cost and relatively easy to adapt for specific use cases.

Massive Open Online Courses

Another opportunity for learners to acquire data science and data analytics skills is through Massive Open Online Courses (MOOCs), which are fully online courses that can support thousands of learners at a time. MOOCs are often free, open-access courses where learners do not typically have to apply or complete prerequisites to enroll. The learners are self-directed and do not engage with the instructor. While course completion does not often lead to a certificate, learners may have the option to pay their MOOC provider for a verified certificate or digital badge indicating the successful completion of the course. Well-known MOOC providers include Coursera, Udacity, Khan Academy, and edX (see, for example, Ngo 2020).

The challenges of using MOOCs in defense domains include concerns about security, rigor, and the reliability of the information provided. Students must also self-regulate and set their own goals. However, several DoD training institutions have partnered with select MOOC providers to deliver training or course content that is readily available, rather than spending months developing new courses. Through these established partnerships, MOOC courses can count toward certificates or digital patches to verify proficiency in various skills and competencies.

Executive and Leadership Training

The growing demand for data analytics and data science skills has seen a concomitant demand for team leaders and decision makers who can facilitate collaboration between technical and business personnel, manage data science teams, communicate relevant data findings, and apply data to decision making. Transforming a team or organization to become increasingly data-centric is a complex endeavor and leaders play a critical role in supporting that transformation. Beyond learning skills specific to the data life cycle associated with their role, non-data skills and strategies can help leaders advance the use of data on their team.

While there is no clear agreement on the non-data skills necessary for team leaders and decision makers, the National Association of Corporate Directors does offer some research-based principles to advance the oversight of digital transformations (Van der Oord et al. 2019). These principles include approaching technology as a strategic imperative, developing continuous technology learning goals and development paths, realigning leadership to reflect the growing significance of technology, and demanding frequent reporting on technology initiatives. While not specific to data use, learning how to apply these or similar principles could help acquisition leaders guide their teams through effective data use.

There are a variety of training programs in data analytics and data science for non-technical managers and leaders throughout higher education and the commercial sectors, including certificate programs. Data science programs for leaders may offer opportunities to learn both data and non-data skills, and many utilize a blend of lectures, interactive discussions, and case studies. Content can be delivered either in person or virtually, and costs can range based on the content and training provider. Additionally, some organizations partner with higher education institutions or third-party providers to develop executive leadership programs with content applicable to the leaders' needs. However, it is important to note that assessments and evaluations of these programs are scarce. As leaders consider training options for their data needs, it is key that they identify programs that align with the skills associated with their data roles.

Conclusion 6.1: Higher education and other training are at an inflection point in their delivery of online education and training; changes are happening rapidly. It is possible to monitor these developments to take advantage of the new opportunities that higher education offers through training partnerships to provide additional data analytics and data science training for acquisition personnel.

DATA USE AND ANALYSIS TRAINING OPPORTUNITIES FOR DEFENSE ACQUISITION PERSONNEL—CURRENT PROGRAMS

Personnel in the defense acquisition workforce, as well as a large proportion of support contractors, receive training for their jobs through Defense Acquisition University (DAU). The mission of DAU is to "provide for—(1) the professional educational development and training of the acquisition workforce; and (2) research and analysis of defense acquisition policy issues from an academic perspective" (10 USC, Sec. 1746). DAU has several "brick and mortar" campuses across the United States, but many of its courses and training programs are provided online. The institution both develops its own courses and works with external vendors to offer a broad selection of courses relevant to the defense acquisition workforce.

Within defense acquisition, personnel are certified for their respective positions. DAU provides certification across 14 different career fields, and within each career field, there are three tiers of certification. Based on testimony provided to the committee in October 2018 by Ms. Megan McKernan from the RAND Corporation, DAU offers more than 220 courses related to data and/or analytics, and enrollments in these courses topped 200,000 in the 2018 fiscal year. In addition, DAU offered a webinar series in 2020 on digital readiness that covered some fundamentals of data science (DAU 2020a). The learning outcomes for these courses are unclear, but the time frame for completion is often brief. Ms. Darlene Urquhart, Director of the Enterprise Integration Directorate at DAU, informed the committee during its December 2019 meeting that DAU was interested in training learners to identify problems that data analytics could help them solve. However, Urquhart shared that DAU had not yet hired a data scientist to its faculty and there is an institutional preference to utilize commercially available training for data analytics training, such as MOOCs.

A site visit to DAU in January 2020 provided insights into the institution's current and future efforts to meet data analytics and data science needs for acquisition personnel. During the site visit, committee members heard from DAU leadership and interacted with faculty and learners in a

300-level program management course and a 400-level program manager case study course. When asked about training needs in data analytics, learners in the 300-level classes expressed their interest in learning how to ask "the right questions," find relevant data, and use the appropriate skills for data analysis. However, trainees shared that the data they require are sometimes inaccessible or have been manually entered into basic spreadsheets, making the data difficult to manipulate, analyze, and verify.

During the classroom visits, several committee members also observed that the data sets used in case studies and computer simulations incorporated oversimplified data, and learners did not need to address uncertainty in their decision making. Uncertainty relates to the understanding that while data are rarely perfect, when properly analyzed they can be sufficient for guiding decision making. It is important that people who make decisions based on data recognize that real data are often messy, variable, and represent only a sample. Learners are often surprised by how much effort is involved in finding and cleaning data. Similarly, when they encounter real data sets in the workplace, there is concern about ambiguity because they did not encounter it in their training. For instructors with some level of data acumen or data literacy, incorporating realistic data sets into courses and discussing the value and complexity of those data may enable learners to better integrate the methods from class into their daily work (American Statistical Association 2016).

At both the site visit and the committee's data-gathering workshop, the Director of User Experience and Platform Optimization at DAU, Ms. Maryann Watson, shared that high interest in data and analytics courses drove efforts at DAU to develop an optional credentialing program in October 2019 to deepen skills beyond its three-tier certification program. The credentialing program provides acquisition personnel with targeted and job-specific learning experiences. Some DAU credentialing courses, such as for agile software development and data analytics, are offered by commercial MOOC vendors such as Coursera (Pearson and Trevino 2019). While current credentialing courses are focused on entry-level workers, DAU anticipates developing more advanced credentials in the future. As of summer 2020, DAU offered the certification program Data Analytics for DoD Acquisition Managers.

As DAU expands data analytics training through its new credentialing pathway, the Naval Postgraduate School (NPS) and the Air Force Institute of Technology (AFIT) offer graduate-level programs in data analytics that are open to acquisition personnel. In addition to their graduate offerings, both institutions also provide continuing education opportunities and formal certificates where acquisition personnel can access training in data analytics.

NPS offers a four-course certificate in data science that includes a blend of lecture and lab hours. The objective of the NPS data science certificate program is to provide learners with training in statistical and machine learning techniques to manage and gain insights from data. Learners enrolled in the program are required to have some background in statistics and programming languages.

NPS also offers a certificate program in Data Sciences and Analytic Management. Dr. William Muir, Assistant Professor in the Graduate School of Defense Management at NPS, informed the committee during its October 2019 meeting that one of the goals of the data science and analytics programs is to train NPS graduates for effective collaboration among data analysis, scientists, statisticians, and leaders. NPS faculty incorporate publicly available and relevant data into their courses, both within and outside the certificate program in Data Sciences and Analytic Management. As a result, NPS students can access real-time acquisition data, including streaming data, which emphasizes the need for the interoperability of data and data sharing. Muir shared that there are data sets that can take an acquisition officer between 6 and 12 months to acquire, but that same data can be accessed by NPS students in fewer than 15 minutes. There is potential for downstream frustration when graduates re-engage in their acquisition projects and are unable to readily access relevant data at the speed NPS provided.

AFIT offers a certificate in data analytics and another in data science. The data analytics certificate program is a five-course distance learning experience that includes introduction to data analytics, data and databases, introduction to machine learning, statistics, and computer programming with Python. The certificate program in data analytics is focused on the use and understanding of data applications and tools, not mathematical theory and algorithm development. AFIT's data science certificate program (DSCP) offers graduate-level training on advanced analytic techniques, large and complex data sets, and computer programming. In addition to training learners in all aspects of the data life cycle, the DSCP also provides training on speaking to disparate groups within an organization to implement data science applications and solutions.

Defense acquisition personnel can enroll in data analytics degree programs, certification pathways, and continuing education courses at DoD-affiliated institutes such as DAU, NPS, and AFIT, but there are also opportunities for personnel to receive data training at non-DoD institutions. In some instances, DoD educational institutions engage with external partners to fulfill training criteria. For example, DAU offers an Equivalency Program, which provides an opportunity for higher education institutions, other DoD schools, commercial vendors, and professional associations to offer courses, programs or certifications that DAU will accept as equivalent (DAU 2020b). As of July 2020, at least 16 colleges and universities provided courses approved by DAU's Equivalency Program, but none of the approved courses included data analytics or data science. However, the DAU Equivalency Program could provide opportunities in the future for acquisition personnel to engage in additional training that builds on the new data analytics certification program. Broadening DAU's educational partnerships allows DoD to outsource some training and potentially develop hiring pipelines at partner institutions.

While not specific to defense acquisition, the Army launched a higher education partnership program in 2020 to build personnel capabilities in data science and artificial intelligence. The Army Artificial Intelligence Task Force is a two-year pilot program with Carnegie Mellon University (CMU) providing master's degrees in data science and data engineering to both uniformed and civilian Army personnel (Army Futures Command 2020). Army also is also partnering with CMU to launch a future artificial intelligence executive education program to provide Army leadership with a high-level understanding of data science to make informed and strategic decisions to integrate data capabilities into Army operations (CMU 2020).

At its data-gathering workshop in April 2020, the committee learned about emerging data training efforts across the U.S. Department of the Navy (DON) and the U.S. Department of the Air Force. While not specific to the acquisition workforce, Mr. Thomas Sasala, Chief Data Officer of DON, noted that DON has created partnerships with commercial companies to offer micro-degrees to build up data acumen, and public universities, such as Old Dominion University, to offer on-site training. Similarly, Ms. Eileen Vidrine, Chief Data Officer of the Air Force, shared how the upcoming establishment of the Air Force Digital University (the Digital University) will include block-style short courses on various topics including data use and data science for airmen and Air Force civilian personnel (Kanowitz 2020; Barnett 2020a, 2020b). The Digital University has engaged in partnerships with commercial training companies, such as Udacity and Pluralsight, to deliver technical content and boost digital literacy in the Air Force. Courses in the Digital University will also be designed to provide accessible courses for senior executives and high-ranking uniformed officers.

Other federal efforts are under way to provide training resources for critical data skills. According to Action 13 of the Federal Data Strategy 2020 Action Plan, the General Services Administration (GSA) will develop a curated catalogue of existing training providers, programs, courses, certifications, and other opportunities for federal personnel to practice and apply new data skills (Federal Data Strategy 2020). While not specific to data skills, DoD's Enterprise Course Catalog will centralize tens of thousands of training and course offerings so that they are searchable through a single web portal (Advanced Distributed Learning Initiative 2020). These catalogues could be useful tools in the future for accessing data training resources and developing data training plans.

An emerging DoD acquisition training program for artificial intelligence (AI) could reveal additional strategies for training acquisition personnel in data science and data analytics. In October 2020, the AI Education Strategy was piloted with 84 DoD acquisition and requirements personnel to increase AI capabilities (DoD 2020). Key to the AI Education Strategy are workforce archetypes with specific AI roles and training needs. Each AI archetype has subcategories of learning outcomes and competencies tiered by levels of proficiency that connect with curricular recommendations. Close monitoring of this program and the utility of the archetypes could be informative to future data science and data analysis training efforts for defense acquisition personnel.

Overall, DoD already has access to a variety of data training options and partnerships that could potentially be expanded for defense acquisition personnel. Opportunities will continue to evolve in response to demand for training in data science and data analytics, but it is key that as programs are developed and adapted for acquisition personnel, they are assessed and evaluated for their success in increasing learner competencies in data use for acquisition contexts. The committee encourages DAU, NPS, and AFIT to continue piloting new approaches to delivering data analytics and data science training to acquisition personnel, but evaluations and assessments are essential to those piloting efforts. As the acquisition workforce engages in new data training programs, clear evaluation plans will help inform future efforts for scaling and modifying the programs to best address acquisition data needs.

> **Finding 6.1:** Some data-relevant courses in DoD use oversimplified data or scenarios with limited access to real data, preventing the student from learning how to deal with the inherent uncertainties and challenges in real acquisition data with all their complexity and quality issues.

> **Recommendation 6.1:** Institutions that provide training for defense acquisition professionals should ensure that courses integrate realistic data and challenges in currently available courses, including non-data-focused courses. Realistic data and challenges offer students the opportunity to learn about uncertainty, sampling, variability, and noise. This realism provides personnel opportunities to learn and apply data techniques in acquisition scenarios and projects.

> **Recommendation 6.2:** The defense acquisition system should continue to leverage and expand the variety of data science training options available both within and outside of the Department of Defense for the acquisition workforce. Trends include higher education, certificate

programs, data "bootcamps," micro-credentialing, online courses, and executive and leadership training. These options include data analytics and data science training programs offered by the Military Services, internal and external higher education institutions, and the commercial sector.

Finding 6.2: Evaluations and assessments are essential to pilot training efforts.

Recommendation 6.3: New training programs for acquisition leadership and personnel in data science should incorporate assessment parameters to evaluate their success. Successful approaches and programs can be expanded to increase effectiveness and broaden the data capabilities of the defense acquisition workforce.

INDUSTRY AND GOVERNMENT TRAINING APPROACHES IN DATA SCIENCE AND ANALYTICS

Given the competition to hire data talent, many organizations across the United States are addressing their increased demand for data capabilities by developing their own training programs, pathways, or partnerships. While not all the non-DoD training programs reviewed by the committee are directly applicable to the defense acquisition experiences, there are opportunities to apply some industry and government training strategies.

On a virtual site visit to Lockheed Martin, Mr. Bruce Litchfield (Vice President, Sustainment Operations at Lockheed Martin) shared that increasing data use at such a large company required a cultural transformation. While technological capabilities are critical for effective data use, success is tied to leadership, people, and processes. Lockheed Martin started with clarifying its vision to align with its desired outcomes: aiming to build a data-centric enterprise that collects, integrates, and analyzes data to continually improve performance. Once the data vision was established, transformation efforts at Lockheed Martin focused on building a system that addressed the data needs most relevant to its vision. To improve data literacy for approximately 50,000 employees at Lockheed Martin, the company selected the basic data skills its employees need to add value to the corporation while not detracting from processes and policies already in place. A catalogue of training modules was developed around the selected skills and include many free or low-cost modules offered by commercial vendors. Employees then take a data literacy assessment to inform them on which training the individuals should complete. A more intense training program is offered for Lockheed Martin employees working toward the position of data scientist or data engineering. While the specifics of this training program were not shared at the site visit, approximately 1,000 out of approximately 50,000 Lockheed Martin employees participate in the program.

Mr. Melvin Greer, Chief Data Scientist with Intel USA and committee member, discussed at the data-gathering workshop how Intel approaches its goal of training all its employees in data science. While Intel is known as a semiconductor manufacturer, it has pivoted its focus to data analysis. Intel has identified the development, growth, and analysis of data as central to its ability to innovate and grow. As a result, all employees are provided training in data science and AI. Intel's training was developed internally, and the curriculum provides a foundational understanding of data science. Intel's required training is not intended to develop every employee into a data engineer or analyst, but to instead ensure that every employee has a basic understanding of how data science contributes to Intel's development and growth as a data-centric organization.

Federal agencies outside of DoD also offer personnel opportunities for data training, and training material and structure may be scalable to DoD-led efforts. For example, a program known as Commerce Academy at the Department of Commerce piloted freely available courses, including a course on data science for leadership. When the Commerce Academy was running, a high level of interest was reported in both the interactive and recorded courses. No evaluation of the program was conducted, but former Commerce Academy director, Natassja Linzau, shared her reflections of the strengths and weaknesses of the training program. According to Linzau, the training courses were most beneficial when Commerce Academy staff directly provided the training because they could offer contextual framing of the learning materials through relevant case studies. The training program offered multiple learning formats, including in-person and synchronous virtual courses, to accommodate various personnel schedules. However, the Commerce Academy staff encountered difficulties providing necessary software on learners' computers, utilizing a viable set of relevant data in the training modules, and sustaining the program through staff transitions.

In response to increasing data collection, the U.S. Department of Health and Human Services (HHS) developed the Data Science CoLab to build a community of employees as skilled data scientists.[2] A cohort from HHS participates in an eight-week program that includes a blend of courses in data science, statistics, and programming, and concludes with a capstone project. Courses are taught by external experts, but the cost of the program is fully covered by HHS. As of 2021, the Data Science CoLab is a fully

---

[2] HHS, "The HHS Data Science CoLab," last reviewed on August 14, 2020, https://www.hhs.gov/cto/initiatives/data-science-colab/index.html.

virtual training experience and accessible to contractors as well as HHS employees.

Some industry organizations invest in partnerships with higher education institutions to offer learners, or potential employees, opportunities to hone their training in data analytics or data science through real-time application. For example, the University of Massachusetts at Amherst offers a summer program that matches data science undergraduate and graduate students to MassMutual in order to work on external projects. MassMutual also worked with the university to establish a college of data science and a master's degree program. In this scenario, the higher education institution serves as a talent pool for the company, and MassMutual can hire graduates of the program with foundational skills in data science.

ENVISIONED FUTURE FOR TRAINING IN DATA USE CAPABILITIES FOR DEFENSE ACQUISITION PERSONNEL

Defense acquisition handles large volumes of data across different tools and platforms. Personnel need to be empowered to harness those data. As outlined in this chapter, there are a variety of training programs to increase data capabilities for the defense acquisition workforce so they can use that data to inform better decision making. Of the programs listed in this chapter, there is no one-size-fits-all training approach. Rather, acquisition teams can consider their current and projected data needs and select training programs that enable them to meet those needs. Training requirements will shift depending on professional roles, the application of data, the scale of data, and mission objectives. Certain positions may require more domain-expertise while others are more data-centric, but importantly, high levels of both domain and data science expertise for an acquisition personnel are unreasonable (Anton et al. 2019). An effective data training strategy will involve coordination across teams to balance data-related capabilities with acquisition-specific expertise.

Training and sustaining a data-capable workforce in defense acquisition are essential and require resources and leadership, especially as the field of data science continues to evolve and advance. To keep up with these advancements, training programs for the acquisition workforce will need to continually adapt while being accessible to learners of all backgrounds with limited available time and resources. Providing a variety of training opportunities for increasing the data capabilities of the defense acquisition workforce will enable data-informed decision making to improve cost, schedule, and performance across DoD.

Recommendation 6.5: Defense acquisition leadership should take program goals and characteristics into account when selecting or developing their data science training programs. Clarifying the data goals and key data capabilities necessary for different acquisition teams are essential steps for identifying suitable training programs.

Recommendation 6.6: Collectively, these recommendations should be considered by the defense acquisition leadership as part of the larger Department of Defense (DoD) Data Strategy. While DoD has a data strategy, it lacks specific guidance for how data science can improve defense acquisition. Specific guidance should address not only phases of the data life cycle and consumption, but also the workforce that facilitates these functions and is central to the data life cycle. Tradeoffs and investment limitations abound, so a strategic plan is critical to guiding and ensuring prioritized investments to maximize payoff.

REFERENCES

Advanced Distributed Learning Initiative. 2020. "Enterprise Course Catalog (ECC)." https://adlnet.gov/projects/ecc/.
American Statistical Association. 2016. "Guidelines for Assessment and Instruction in Statistics Education (GAISE) in Statistics Education (GAISE) - College Report 2016." https://www.amstat.org/asa/files/pdfs/GAISE/GaiseCollege_Full.pdf.
Anton, P.S., M. McKernan, K. Munson, J.G. Kallimani, A. Levedahl, I. Blickstein, J.A. Drezner, and S. Newberry. 2019. Assessing Department of Defense Use of Data Analytics and Enabling Data Management to Improve Acquisition Outcomes, Santa Monica, CA: RAND Corporation, RR-3136-OSD, August. https://www.rand.org/pubs/research_reports/RR3136.html.
Army Futures Command. 2020. "Artificial Intelligence Task Force Welcomes Inaugural Class of AI Scholars." U.S. Army. https://www.army.mil/article/237258/artificial_intelligence_task_force_welcomes_inaugural_class_of_ai_scholars.
Barnett, J. 2020a. "As Air Force's Digital U Grows Its Ranks, It Looks to Refine Course Work." FedScoop. https://www.fedscoop.com/air-forces-digital-u-number-of-users-subject-matter-experts/.
Barnett, J. 2020b. "Air Force's Digital University Prepares for Launch with a Focus on 'Tactical Operators'." FedScoop. https://www.fedscoop.com/air-forces-digital-university-free-technical-training/.
CMU (Carnegie Mellon University). 2020. "CMU Partners with U.S. Army To Grow Data Science and AI Expertise." https://www.cmu.edu/news/stories/archives/2020/september/army-partners-grow-data-science.html.
DAU (Defense Acquisition University). 2020a. "Explore the Current Series of Webcasts." Last updated https://www.dau.edu/dau-webcasts/p/Explore-Webcast-Series.
DAU. 2020b. "Defense Acquisition University Equivalency Program." Defense Acquisition University. Last updated https://icatalog.dau.edu/appg.aspx.
DoD (Department of Defense). 2020. "DoD Data Strategy." https://media.defense.gov/2020/Oct/08/2002514180/-1/-1/0/DOD-DATA-STRATEGY.PDF.
Federal Data Strategy. 2020. "Federal Data Strategy: 2020 Action Plan." https://strategy.data.gov/action-plan/#action-13-develop-a-curated-data-skills-catalog.
Kanowitz, S. 2020. "Is the AF's Digital University the Future of IT Training?" Defense Systems. https://defensesystems.com/articles/2020/10/21/air-force-digital-university.aspx.

Ngo, C. 2020. "10 Best Free and Affordable Platforms for Online Courses," Fort Mill, SC: Best Colleges, April. https://www.bestcolleges.com/blog/platforms-for-online-courses/.

Pearson, D., and A. Trevino. 2019. "Refreshing Acquisition Workforce Skills." Defense Acquisition University. https://www.dau.edu/library/defense-atl/blog/Refreshing-Acquisition--Workforce-Skills.

Van der Oord, F., S. Mezeu, L. Chacko, and R. Lam. 2019. *Governing Digital Transformation and Emerging Technologies: A Practical Guide*. The National Association of Corporate Directors.

7

Findings, Conclusions, and Recommendations

The findings, conclusions, and recommendations from the prior chapters in this report are provided below to provide a concise review of the major points from the committee's study.

Finding 1.1: By making decisions informed by data, leadership sets an example that can promote the cultural changes necessary to facilitate data-informed decisions across the entire department.

Recommendation 1.1: The Department of Defense (DoD) and congressional leaders and stakeholders should promote the cultural changes necessary to facilitate data-informed decisions across the entire DoD by insisting that high-quality, complete, and accurate data and analysis be provided to inform their own decisions and those of their subordinates.

DATA SCIENCE IN DOD ACQUISITION

Finding 4.1: Data science has improved acquisitions processes by enhancing program and contracting analysis, tracking cost and program performance, enabling analysis of alternatives, and informing decision making.

Finding 4.2: There are several current shortcomings in the front end of data science processes in defense acquisition—namely in, data collection and curation.

Finding 4.3: Data silos, wherein data are not shared across DoD, are a common obstacle preventing the utilization of the full potential of the data life cycle.

Conclusion 4.1: With support from defense acquisition leadership, improved data-sharing policies and support for front-end phases of the data life cycle are major short-term opportunities to improve acquisition processes.

Conclusion 4.2: A common data infrastructure and platforms, personnel training and tools, and access to DoD-wide data are major long-term data science investments that will have long-term positive impact on acquisition processes.

Conclusion 4.3: Improved data use and the data life cycle could be applied to urgent, high-profile challenges in defense acquisition. Doing so would serve as exemplars of the value of data use and data-informed decision making in acquisition, serving as a catalyst for further change.

Recommendation 4.1: The Department of Defense should continue to seek improvements in defense acquisition through the increased application of data science, including addressing shortfalls in data collection, curation, management, and sharing.

DATA LIFE CYCLE MINDSET, SKILLSET, TOOLSET: ROLES AND TEAMS

Finding 5.1: An ever-increasing majority of the acquisition workforce is participating in the data life cycle. Just as defense acquisition is most successful when it is accomplished collaboratively, so is data science. Data science is an inherently collaborative field where success requires cooperation, communication, and coordination. Phases of the data life cycle directly or indirectly depend on one another, and those participating in the data life cycle must be able to relay key details to others.

Conclusion 5.1: Defense acquisition workforce members' awareness of their value and individual role in the data life cycle is critical for data science applications. Barriers to this awareness include (1) lack of familiarity and use of the data life cycle, (2) adoption challenges for data science technologies and practices, and (3) widespread self-doubt of their own skillsets and insecurity or anxiety with STEM (generally) and data science (in particular).

Conclusion 5.2: Improved data use in acquisition requires a collective skillset and teamwork among the acquisition workforce members. Team members will need data skills appropriate for their functional role in the acquisition community.

Finding 5.2: Upon graduation, undergraduate students are increasingly expected to have learned and applied some level of data science knowledge and skills.

Conclusion 5.3: Defense acquisition can expect the future workforce to have a baseline set of data science skills.

Finding 5.3: While specific outcomes are being updated and refined, *data literacy* for many undergraduate students includes an understanding of the data life cycle and how it works, data story-telling and communication, and an ability to recognize matters of ethics, privacy, and security.

Recommendation 5.1: The Department of Defense and its components should ensure that all members of the acquisition workforce and its leadership eventually have a common baseline data literacy, which includes an understanding of the data life cycle and how it works, data story-telling and communication, and an ability to address matters of data ethics, privacy, and security—all skills that may evolve along with the use of data science in government, industry, and academia.

Finding 5.4: Data scientists are experts across the data life cycle, with special emphasis on advanced techniques for collection, curation, management, analysis, and visualization.

Finding 5.5: Due to an evolving data science curriculum in higher education, not all data science degrees prepare students with the same skillsets.

Finding 5.6: Executing the data life cycle is a collaborative endeavor and generally requires a collective skillset found in teams of data engineers, data scientists, data analysts, data users, domain experts, and leaders/decision makers.

Conclusion 5.4: A data-capable defense acquisition workforce must—as a whole—have access to the full range of data science capabilities, from data collection and curation to data analysis and visualization.

Conclusion 5.5: Management of data science projects uses strategies and approaches for leading collaborative, cross-functional, technical projects with specific attention paid to the development of a team that has skills across the data life cycle and to asking questions specific to the quality and utility of data.

Recommendation 5.2: The Department of Defense should prioritize the utilization of data and the data life cycle by appropriate and judicious investment in the data science mindset, skillset, and toolset of the acquisition workforce.

PREPARING AND SUSTAINING A DATA-CAPABLE WORKFORCE

Conclusion 6.1: Higher education and other training are at an inflection point in their delivery of online education and training; changes are happening rapidly. It is possible to monitor these developments to take advantage of the new opportunities that higher education offers through training partnerships to provide additional data analytics and data science training for acquisition personnel.

Finding 6.1: Some data-relevant courses in DoD use oversimplified data or scenarios with limited access to real data, preventing the student from learning how to deal with the inherent uncertainties and challenges in real acquisition data with all their complexity and quality issues.

Recommendation 6.1: Institutions that provide training for defense acquisition professionals should ensure that courses integrate realistic data and challenges in currently available courses, including non-data-focused courses. Realistic data and challenges offer students the opportunity to learn about uncertainty, sampling, variability, and noise. This realism provides personnel opportunities to learn and apply data techniques in acquisition scenarios and projects.

Recommendation 6.2: The defense acquisition system should continue to leverage and expand the variety of data science training options available both within and outside of the Department of Defense for the acquisition workforce. Trends include higher education, certificate programs, data "bootcamps," micro-credentialing, online courses, and executive and leadership training. These options include data analytics and data science training programs offered by the Military Services, internal and external higher-education institutions, and the commercial sector.

Finding 6.2: Evaluations and assessments are essential to pilot training efforts.

Recommendation 6.3: New training programs for acquisition leadership and personnel in data science should incorporate assessment parameters to evaluate their success. Successful approaches and programs can be expanded to increase effectiveness and broaden the data capabilities of the defense acquisition workforce.

Recommendation 6.5: Defense acquisition leadership should take program goals and characteristics into account when selecting or developing their data science training programs. Clarifying the data goals and key data capabilities necessary for different acquisition teams are essential steps for identifying suitable training programs.

Recommendation 6.6: Collectively, these recommendations should be considered by the defense acquisition leadership as part of the larger Department of Defense (DoD) Data Strategy. While DoD has a data strategy, it lacks specific guidance for how data science can improve defense acquisition. Specific guidance should address not only phases of the data life cycle and consumption, but also the workforce that facilitates these functions and is central to the data life cycle. Tradeoffs and investment limitations abound, so a strategic plan is critical to guiding and ensuring prioritized investments to maximize payoff.

SYNOPSIS

Based on testimonies and information collected as well as its experience and expertise, the committee believes that there are viable and cost-effective opportunities for DoD to improve defense acquisition processes and decision making through expanded use of the evolving capabilities in data use and data science broadly. Exactly what areas and what level of investment are appropriate given competing demands and other opportunities for the department must be decided by a strategic review, assessment, and plan of data-science needs, opportunities, costs, and anticipated returns against the return-on-investment from other opportunities. This report can be used—in guiding such a review, assessment, and plan—to embed the data life cycle into the acquisition process, establish roles and teams for executing the data life cycle, and build training opportunities for the acquisition workforce. Attention to workforce empowerment and training throughout the data life cycle, as outlined in this report, will further enhance DoD efforts to improve data use in defense acquisition, incorporate data-informed decision making, and advance the acquisition process.

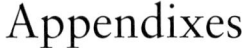

A

Meeting and Workshop Agendas

FIRST COMMITTEE MEETING
WASHINGTON, D.C.
OCTOBER 28, 2019

13:00-14:45 **Perspectives from Study Sponsor**

    Ms. Stacy Cummings, Principal Deputy Assistant Secretary of Defense, Acquisition Enablers, Department of Defense

    Mr. Mark Krzysko, Deputy Director of Enterprise Information for the Office of the Under Secretary of Defense for Acquisition, Technology, and Logistics Acquisition Resources and Analysis, Department of Defense

14:45-15:15 Break

15:15-16:00 **Understanding Data Training at Military Colleges and Universities**

    Dr. Bill Muir, Assistant Professor, Graduate School of Defense Management, Naval Postgraduate School

16:00-17:00   Summary of RAND Analyses on Data Use for Defense Acquisition

          Ms. Megan McKernan, Senior Defense Researchers, RAND Corporation

SECOND COMMITTEE MEETING
WASHINGTON, D.C.
DECEMBER 12, 2019

09:30-10:45   Upskilling Needs and Challenges in DoD Acquisition

          Ms. Stacy Bostjanick, Director, Cybersecurity Maturity Model Certification, Office of the Under Secretary of Defense for Acquisition and Sustainment
          Ms. Rebecca Weirick, Senior Services Manager and Civilian Deputy, Office of the Deputy Assistant Secretary of the Army (Procurement)

10:45-11:00   Break

11:00-12:30   Landscape of Data, Upskilling, and the Acquisition Workforce

          Dr. Yisroel Brumer, Principal Deputy, Cost Assessment and Program Evaluation (CAPE), Deputy Secretary of Defense
          Mr. Garry Shafovaloff, Acquisition and Sustainment Director, Human Capital Initiatives (HCI)
          Dr. Darlene Urquhart, Director, Enterprise Integration Directorate, Defense Acquisition University (DAU)

DEFENSE ACQUISITION UNIVERSITY (DAU) SITE VISIT
FORT BELVOIR, VA
JANUARY 29, 2020

13:00-13:30   Welcome, Introductions, and a Brief Overview of DAU

          Frank L. Kelley, Vice President, Defense Acquisition University
          Darlene Urquhart, Director, Enterprise Integration Directorate and Executive Coach, DAU

13:30-13:45   Break and Relocation to Classroom

13:45-14:30 **Visit PMT 360 and PMT 401**

                 Vance Gilstrap, Director, Major Defense Acquisition Programs, DAU

14:30-14:45 **Foundational Learning (aka classroom and online training)**

                 William Parker, Director, Foundational Learning Directorate, DAU

14:45-15:00 **Workflow Learning (data tools on website, etc.)**

                 Hans Jerrell, Associate Director (Plans & Programs) of our DAU Workflow Learning Directorate, DAU

15:00-15:20 **Performance Learning (aka consulting)**

                 David Robinson, Director, Contractor Cost Data Report Project Office, DAU
                 Carol Tisone, Dean, Mid-Atlantic Region, DAU

15:20-15:30 Break

15:30-16:15 **Transformation, Q&A, and Wrap-Up**

                 Darlene Urquhart, Director, Enterprise Integration Directorate and Executive Coach, DAU

THIRD COMMITTEE MEETING
WASHINGTON, D.C.
JANUARY 30-31, 2020

January 30, 2020

09:00-09:40 **DAU Site Visit Summary**

11:00-12:00 **Open Session with Dr. J. Michael Gilmore**

                 Dr. J. Michael Gilmore, Researcher at the Institute for Defense Analyses, Former Director of Operational Test and Evaluation for the Department of Defense

12:00-13:30 Working Lunch

January 31, 2020

10:00-11:00 **Open Session with Hon. Heidi Shyu**

Hon. Heidi Shyu, Director, VK Integrated Systems, Former Assistant Secretary of the Army for Acquisition, Logistics, & Technology

11:00-12:00 **Open Session with Dr. Sallie Ann Keller**

Dr. Sallie Ann Keller, Division Director, Social and Decision Analytics at the University of Virginia

FOURTH COMMITTEE MEETING
WASHINGTON, D.C.
MARCH 5, 2020

10:00-11:00 **Open Session with Dr. Nancy Spruill**

Dr. Nancy Spruill, Business Acquisition Workforce Functional Leader, Under Secretary of Defense for Acquisition and Sustainment (ret.)

12:00-13:00 **Open Session with Ms. Sezin Palmer**

Ms. Sezin Palmer, Mission Area Executive for National Health, Johns Hopkins University Applied Physics Laboratory

16:00-18:00 **Open Session with Ms. Bethany Blakey and Ms. Bess Dopkeen**

Ms. Bethany Blakey, Director of Change Management within the Office of Governmentwide Policy at the General Services Administration
Ms. Bess Dopkeen, Professional Staff Member at House Committee on Armed Services.

APPENDIX A
WRIGHT-PATTERSON AFB SITE VISIT—AIR FORCE INSTALLATION CONTRACTING CENTER (AFICC)- BUSINESS INTELLIGENCE BRANCH
HELD VIRTUALLY
19 MARCH 2020-1300-1700 EST

AFICC Planned Participants

Mr. Roger Westermeyer-Director, Enterprise Solutions Directorate
Mr. Steve Brady-Director, Enterprise Innovation Division
Mr. Pete Herrmann-Chief, Business Intelligence Branch
Mr. Darin Ashley-Lead, Operations Research Analyst
Ms. Mary Hauber-Lead, Air Force Business Intelligence Tool Database Manager

AGENDA

1. AFICC Mission
2. Organizational Structure and How Business Intelligence Fits into Strategic, Operational and Tactical Levels
3. Dashboard/Deliverable Demonstration of Business Intelligence Using "Operational Data"
4. Data as an Asset-Importance of Investment
5. Basic KSAs Needed
6. Manpower, Team Structure, Incentive
7. Available BI Tools
8. Open Discussion with Leadership and Analysts

WORKSHOP ON IMPROVING DEFENSE ACQUISITION WORKFORCE CAPABILITY IN DATA USE
HELD VIRTUALLY
APRIL 14, 2020

10:00-10:15 **Welcome Remarks and Introduction**

Dr. Tyler Kloefkorn and Dr. Lida Beninson, Co-study Directors
Lieutenant General Wendy Masiello and Dr. Rebecca Nugent, Co-study Chairs

10:15-10:45 **Opening Remarks**

 Mr. Mark Krzysko, Principal Deputy Director, Acquisition Policy and Analytics
 Mr. David Cadman, Acting Principal Deputy Assistant Secretary of Defense, Acquisition Enablers
 Mr. Michael Conlin, Chief Data Officer, U.S. Department of Defense

10:45-10:50 Break

10:50-12:00 **Panel 1—Data Use for Defense Acquisition**

 Mr. Gary Bliss, Director, Performance Assessments and Root Cause, Department of Defense
 Ms. Jennifer Bowles, Director, Land and Naval Warfare Cost Analysis Division, OSD CAPE
 Ms. Lisa Disbrow, Senior Fellow, Johns Hopkins Applied Physics Laboratory
 Lt. Gen. (Ret.) Bruce Litchfield, Vice President of Sustainment Operations, Lockheed Martin Aeronautics
 The Hon. Christine Fox (moderator), Committee Member

12:00-13:00 Break

13:00-14:00 **Panel 2—Perspectives from the Chief Data Officers**

 Mr. Thomas Sasala, Chief Data Officer, US Navy
 Ms. Eileen Vidrine, Chief Data Officer, US Air Force
 Dr. Alyson Wilson (moderator), Committee Member

14:00-14:15 Break

14:15-15:30 **Panel 3—Upskilling Data Capabilities in the Industry Workforce**

 Mr. Melvin Greer, Chief Data Scientist, Intel Corporation Americas
 Dr. Sears Merritt, Head of Data, Strategy and Architecture, MassMutual
 Mr. Paul Nielsen, VP of Strategic Programs, Advanced Technology Collaborative, Optum Technologies

APPENDIX A   *91*

                  Mr. Luis Stevens, Senior Director for High Performance Computing, Target Corporation
                  Dr. Ann McKenna (moderator), Committee Member

15:30-16:00 Break

16:00-17:15 **Panel 4—Educating and Building Data Science Teams**

                  Dr. Darryl K. Ahner, Director, OSD Scientific Test and Analysis Techniques Center of Excellence, Director, Center for Operational Analysis, Professor of Operations Research, Air Force Institute of Technology
                  Dr. Matthew Rattigan, Director of Research Programs for the Center for Data Science, University of Massachusetts, Amherst
                  Professor Jaeki Song, Area Coordinator for Information Systems and Quantitative Sciences, Rawls Business School, Texas Tech University
                  Ms. Maryann P. Watson, Associate Dean for Executive, Internat'l & Reqmts Management Programs at Defense Acquisition University
                  Dr. Rebecca Nugent (moderator), Co-chair

17:15-17:30 **Closing Remarks**

17:30   Workshop Adjourns

LOCKHEED MARTIN SITE VISIT
HELD VIRTUALLY
MAY 4, 2020

Introduction
                  Bruce Litchfield, Lockheed Martin

Digital Transformation and Data Strategy
                  Tammy Foster, Lockheed Martin

Digital Thread and Affordability
                  Renee Pasman, Lockheed Martin

Enterprise Data Approach
                  Mike Baylor, Lockheed Martin

Wrap-up
      Bruce Litchfield, Lockheed Martin

Q&A/Discussion with Committee

NAVAL POSTGRADUATE SCHOOL SITE VISIT
HELD VIRTUALLY
MAY 28, 2020

09:00-09:15 Welcome Remarks and Introduction

      Dr. Tyler Kloefkorn, Study Co-Director, the National Academies
      Dr. Lisa Beninson, Study Co-Director, the National Academies
      Lieutenant General (Ret.) Wendy Masiello, Study Co-Chair, Wendy Mas Consulting, LLC
      Dr. Rebecca Nugent, Study Co-Chair, Carnegie Mellon University

09:15-09:30 **The Naval Postgraduate School**

      Dr. Robert Dell, Acting Provost and Academic Dean, Naval Postgraduate School

09:30-10:15 **Defense Acquisition Educational Programs in Defense Management**

      Overview: Graduate School of Defense Management
      Dr. Yu-Chu Chen, Professor of Economics and Associate Dean of Research, Naval Postgraduate School

      Contracting and Acquisition Program Management, Degree and Non-Degree Programs
      Dr. Rene Rendon, Professor of Acquisition Management, Naval Postgraduate School

      Data Analytics for Defense Management Certificate
      Dr. Jesse Cunha, Professor of Economics, Naval Postgraduate School

      Committee Questions

APPENDIX A

10:15-10:30   Break

10:30-11:00   Acquisition Educational Programs in Engineering and Applied Sciences

   Overview: Graduate School of Engineering and Applied Sciences
   Dr. Clyde Scandrett, Dean, Naval Postgraduate School

   Systems Engineering, Degree and Non-Degree Programs
   Dr. Walter Owen, Associate Chair, Department of Systems Engineering, Naval Postgraduate School

   Data Science Research and Applications in Engineering
   Dr. Brij Agrawal, Distinguished Professor, Department of Mechanical and Aerospace Engineering, Naval Postgraduate School
   Dr. James Scrofani, Associate Professor, Department of Electrical and Computer Engineering, Naval Postgraduate School

   Committee Questions

11:00-11:30   Data Science Educational Programs in Operations and Information Sciences

   Overview: Graduate School of Operations and Information Sciences
   Dr. Dan Boger, Dean, Naval Postgraduate School

   Educational Programs in Analytics and Data Science
   Dr. Matthew Carlyle, Chair, Department of Operations Research, Naval Postgraduate School
   Dr. Jonathan Alt, Assistant Professor, Department of Operations Research, Naval Postgraduate School
   Mr. Eric Eckstrand, Research Associate, Department of Operations Research, Naval Postgraduate School

   Committee Questions

11:30-11:50 **Acquisition Research Program**

    Rear Admiral (Ret.) James Greene, Acquisition Chair, Naval Postgraduate School
    Dr. Robert Mortlock, Principal Investigator, Naval Postgraduate School

11:50-12:00 **Closing Remarks**

    Lieutenant General (Ret.) Wendy Masiello, Study Co-Chair, Wendy Mas Consulting, LLC
    Dr. Rebecca Nugent, Study Co-Chair, Carnegie Mellon University

B

Defense Acquisition Notes

This appendix provides a brief overview of the complexity of acquisition in the Department of Defense (DoD) and how that complexity poses challenges for applying data science to defense acquisition. Those challenges are realized in the various categories of data that need to be collected and managed, the domain insight required by analysts, and the complexity in communicating data findings to consumers of the analysis. As with commercial sector, an understanding of both data science and the application domain is necessary for the effective use of data, and this brief introduction illustrates the depth of understanding needed to make meaningful improvements in defense acquisition.

Defense acquisition includes the full life cycle of a system, from conceptualization through development, testing, production, operation, sustainment, and disposal (Box B.1). Thus, acquisition includes not only development and procurement (what some commonly refer to as "acquisition"), but also product support, modification, and disposal of weapon systems.

ACQUIRED ITEMS BY FUNDING AND CONTRACTING CATEGORIES

Financial resources for defense acquisition are categorized in various ways that help Congress control where the resources are spent. Figure B.1 and Box B.2 illustrate how funding levels in different acquisition categories change dramatically over time as budgets evolve over time. These changes hold when adjusted for inflation as well.

BOX B.1
Definition of Defense Acquisition

Defense Acquisition: The conceptualization, initiation, design, development, test, contracting, production, deployment, integrated product support, modification, and disposal of weapons and other systems, supplies, or services (including construction) to satisfy DoD needs, intended for use in, or in support of, military missions.

SOURCE: Department of Defense, 2020. Directive 5135.02. Under Secretary of Defense for Acquisition and Sustainment (USD(A&S)). https://www.esd.whs.mil/Portals/54/Documents/DD/issuances/dodd/513502p.pdf?ver=2020-07-15-133537-053.

Major Accounts:

- **Research, Development, Test, and Evaluation** (RDT&E)
    - Science and Technology (6.1–6.3; basic and applied research, adv. technology)
    - Development   (6.4, 6.5, 6.7)
    - Management   (6.6)
- **Procurement**
- **Operations and Maintenance** (O&M)
    - Operate and sustain people and systems
- **Military Personnel** (MilPer)

Smaller accounts:

- Military Construction (MILCON)
- Family Housing
- Revolving and Management Funds

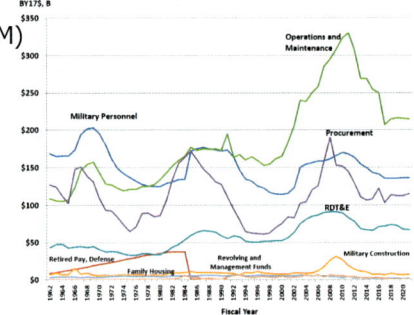

FIGURE B.1 Acquisition financial resources. SOURCE: Data from Under Secretary of Defense, Acquisition, Technology, and Logistics, 2016, *Performance of the Defense Acquisition System: 2016 Annual Report,* Washington, DC: Department of Defense.

One challenge facing acquisition program managers is the management of multiple budgets, including the current budget, the requested budget, the upcoming budget, and the preparation of budgets for future years. Some types of funding can be obligated and expended across multiple years, while others must be obligated within the year it is appropriated by Congress.

The majority of DoD's supplies, equipment, and services are acquired through contracts. Figure B.2 presents a categorization of these contracted acquisitions from fiscal year 2008 through 2015. Importantly, contract data

APPENDIX B

> **BOX B.2**
> **What Is Meant by the "Color of Money"?**
>
> The "color of money" refers to the general spending category where the funding is restricted. Money is authorized and appropriated by Congress to specific accounts. The major accounts are research, development, test, and evaluation (RDT&E); procurement; operations and maintenance (O&M); and military personnel. There are smaller accounts as well, including military construction (MILCON), family housing, and revolving and management funds. Funding for each account flows through defense acquisition at some point.

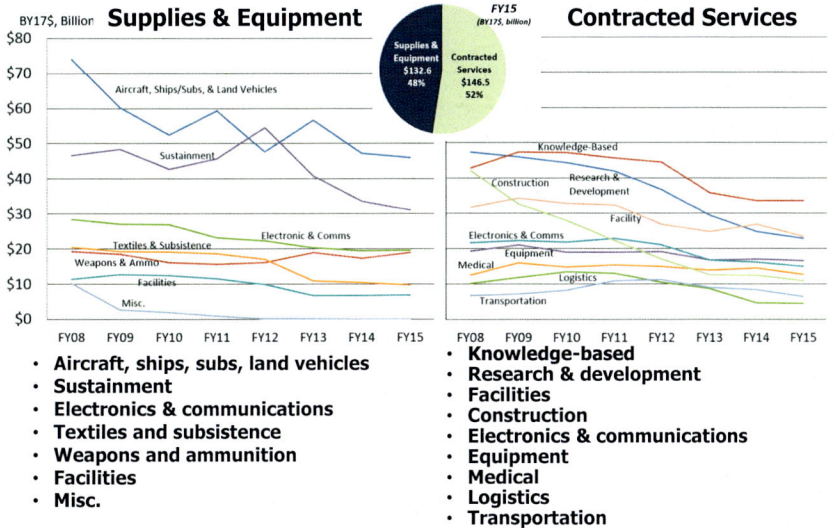

- Aircraft, ships, subs, land vehicles
- Sustainment
- Electronics & communications
- Textiles and subsistence
- Weapons and ammunition
- Facilities
- Misc.

- Knowledge-based
- Research & development
- Facilities
- Construction
- Electronics & communications
- Equipment
- Medical
- Logistics
- Transportation

FIGURE B.2 What does the Department of Defense acquire? SOURCE: Data from Under Secretary of Defense, Acquisition, Technology, and Logistics, 2016, *Performance of the Defense Acquisition System: 2016 Annual Report*, Washington, DC: Department of Defense.

for a single contract can have different line items for acquiring goods or services in different categories, thus complicating data collection and analysis.

ACQUISITION PROGRAMS AND CATEGORIES

Programs are the mechanism by which most major acquisitions are undertaken. The acquisition processes, now referred to as "pathways" in the Adaptive Acquisition Framework, associated with the various programs are tailored to specific types of acquisition (i.e., urgent capability, "middle tier," major capability, software, business systems, and acquired services—see Figure 2.1; (DoD Instruction 5000.02, 2020).

ACQUISITION MANAGEMENT AND OVERSIGHT LEVELS

The use of data and analyses required in different acquisition scenarios varies across management levels and oversight (Figure B.3). Functional, or tactical data, are required for analyses by engineers, contractors, and subcontractors. At more executive levels, only summary data are generally available or used. From a data science perspective, executives generally need high-level summaries and perspectives, but the staff that develop these summaries need to analyze more detailed data.

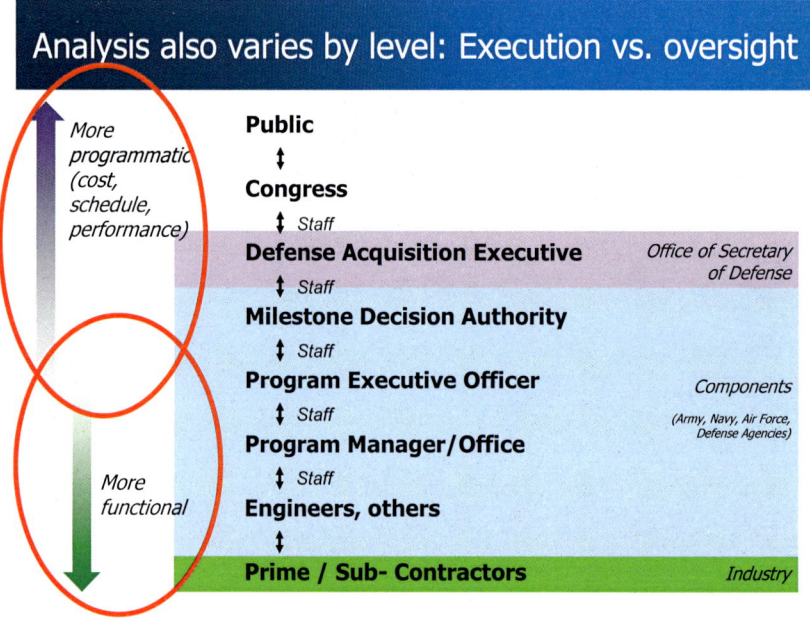

FIGURE B.3 Analysis also varies by level—execution versus oversight.

APPENDIX B

COMMUNITIES THAT AFFECT ACQUISITION

Three primary communities enable defense acquisition: the acquisition community itself (DAS), the budgetary and financial management community (PPBE), and the requirements community (JCIDS). In order to acquire anything, both a budget and a validated requirement is necessary. See Figure B.4.

Each of these three communities has its own unique data processes. DoD and the components have their own data systems to support the acquisition, budget, and requirements processes. While data submitted to the Office of the Secretary of Defense (OSD) must conform to data definitions set by OSD in concert with the military departments, program data within each department's systems have their own data definitions that may not align with seemingly similar data in other departments, which magnifies the complexity of accessing and analyzing data across DoD.

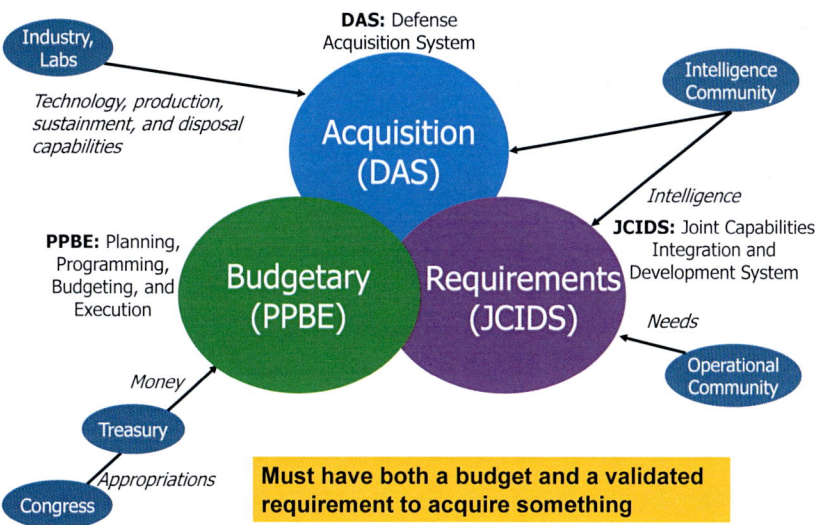

FIGURE B.4 Acquisition involves three interrelated but separate systems and communities.

DEFENSE ACQUISITION FUNCTIONS AND THE DATA LIFE CYCLE

As discussed in Chapter 4, data science is applied to defense acquisition through the range of acquisition functions and processes and ultimately in different domains of decisions and supporting analyses and data. Table B.1 helps to illustrate this richness. The interested reader is referred to Anton et

TABLE B.1 Categories of Acquisition Domain Decisions and the Underlying Data and Analysis

Acquisition Function	Example Decision	Relevant Phases of Data Life Cycle
Acquisition Oversight	Program satisfies larger portfolio needs	**Collect, manage, integrate,** and **analyze** data on current operational systems (commodity, inventory, age, condition, capabilities); simulation, modeling, and **analysis** of missions (e.g., kill chains).
Program Management		
• Business case, economic, and budget analysis	Program affordability	**Collect, manage, integrate,** and **analyze** program-related data on costs, schedule, performance, risk, current market, and alternatives
• Acquisition strategy	Opportunities for cooperation	**Identification** and use of data aligned with decisions related to objectives; **shared** data infrastructure and **access**; analysis on contracting, intellectual property, equipment and architecture, multi-year procurement, risk
Production, Quality, and Manufacturing (PQM)	Production management	**Collect** data on production quality issues (e.g., scrap rates), and manufacturing performance relative to schedule. Management of complex databases; design and **analysis** for engineering, manufacturing, quality, and reliability
Acquisition Interface Functions	Meets access and analysis requirements	**Identifies, collects, curates,** and **stores** data needed for subsequent **analysis** (engineering, mission, wargaming)
Legal Counsel: Request and Act Upon	Program compliant with laws and policies	**Collecting** memorandums of understanding and relevant documentation; automated use of compliance systems; legal and policy **analysis**

SOURCE: Adapted from Figure B.5, from P.S. Anton, M. McKernan, K. Munson, J.G. Kallimani, A. Levedahl, I. Blickstein, J.A. Drezner, and S. Newberry, 2019, *Assessing Department of Defense Use of Data Analytics and Enabling Data Management to Improve Acquisition Outcomes*, RR-3136-OSD, August, Santa Monica, CA: RAND Corporation, https://www.rand.org/pubs/research_reports/RR3136.html. Courtesy of RAND Corporation.

Domain	Decision	Supporting Analyses	DoD AWF	Other Resources	Analytic Capabilities: Analytic Approaches	Key Data: Available	Gaps and Challenges
Require-ments	Design sufficient and meets requirements	Engineering analysis; Requirements and mission analysis	Engr.; PM	DoD labs; UARCs; FFRDCs; Requirements	Engineering and requirements analysis; wargaming	System design; requirements; threat envelops	Programs have requirements documents, but centralized info. system KM/DS archive is not always complete. Threat data may not be current.
	Requirements validated	Requirements and mission analysis; JCIDS Analysis: Functional Area Analysis (FAA), Functional Needs Analysis (FNA) and Functional Solutions Analysis (FSA)	PM; Leadership	Requirements	*(ref: JCIDS outside DAS)*		
Technical	Technology mature	Technical maturity and risk analysis	Engr.; IT; T&E	DoD labs, test centers, and FFRDCs	Risk analysis; Wide range of T&E analytic approaches and infrastructure		
	Market research; technology non-duplicative	Market research	Engr.; PM	DoD labs; UARCs, and FFRDCs	Internet search; technology and requirements analysis	Market and technical literature	Analyst access to PROPIN.
Risk	Risks acceptable given need	Framing assumptions; RMF	Engr.; IT; T&E; PM	DoD labs; UARCs, Test Centers; and FFRDCs	Framing assumptions; risk analysis; RMF	System concepts; risks	FA concept is new and little understood.
Budgets	Program affordable	Affordability analysis; cost analysis (costs within goals)	CE; PM	CAPE; Resource planners (G8, N8, A8); Leadership; FFRDCs	Budgeting systems, spreadsheets, optimization software	Budget elements; Component planning total (TOA); Cost data	Difficulty exists in mapping budget PEs to specific acquisition programs and detailed acquisition tasks

FIGURE B.5 Categories of acquisition domain decisions and the underlying data and analysis.
SOURCE: P.S. Anton, M. McKernan, K. Munson, J.G. Kallimani, A. Levedahl, I. Blickstein, J.A. Drezner, and S. Newberry, 2019, *Assessing Department of Defense Use of Data Analytics and Enabling Data Management to Improve Acquisition Outcomes*, RR-3136-OSD, August, Santa Monica, CA: RAND Corporation. https://www.rand.org/pubs/research_reports/RR3136.html. Courtesy of RAND Corporation.
(figure continues)

Domain	Decision	Supporting Analyses	DoD AWF	Other Resources	Analytic Capabilities		Gaps and Challenges
					Analytic Approaches	Key Data: Available	
Strategy	Acquisition strategy	Acquisition strategizing	PM, contracting, FM, Leadership	Buying commands; FFRDCs	Acquisition strategizing; objectives; competition strategy; benefits analysis; market research; bundling; business strategy (below); cooperative opportunities; general equipment valuation; Industrial-base analysis; intellectual property analysis and strategizing; architecture (e.g., modular open system?); multi-year procurement analysis; risk analysis; Small-Business analysis (SBIR/STTR).	Requirements, system concepts, industrial base and competitive/cooperative opportunities, small-business capabilities and contributors, risks, cost estimates, equipment design and costs, intellectual property rights and options, architectural standards and options, procurement options.	
Tradeoffs	Tradeoffs made and appropriate	Cost, schedule, and performance tradeoff; Risk analysis; Framing assumptions; Budgetary tradeoffs	PM	Service Chief; FFRDCs	Benefit/cost analysis; cost, schedule, and performance relationship modeling; risk analysis		
	Alternatives considered	Analysis of alternatives (AoAs); Market research	PM, Engr., S&T Manager	CAPE; DoD Labs and FFRDCs; Requirements	Operational military M&S; Technology forecasts; tech. maturity models; risk models;	Operational systems, capabilities, missions, and intelligence; requirements; relating systems; Kill chains; Technology and market data; cost data	Analyst access to PROPIN.

FIGURE B.5 Continued
(figure continues)

Domain	Decision	Supporting Analyses	Analytic Capabilities				Gaps and Challenges
			DoD AWF	Other Resources	Analytic Approaches	Key Data: Available	
Cost	Costs estimated and reasonable	Cost analysis: developmental and market analysis	CE; Contracting; PM	SETA support contractors; UARCs; FFRDCs	Cost analysis (spreadsheets; custom tools; BI)	Prior and current system cost data	Analyst access to PROPIN and smaller program prior cost data.
	Cost within goals	Affordability analysis; Cost analysis: developmental and market analysis	PM	FFRDCs; UARCs			
	Funding available	Budgetary analysis (POM and FYDP)	FM; CE; PM; Lifecycle logistics; PM	Resource planners (G8, N8, A8); FM; Leadership			
	Life-cycle sustainment planned and costed	life-cycle sustainment planning and cost estimating	T&E, Lifecycle Logistics, FM, PM	Sustainment and logistics centers; FFRDCs	O&S cost estimating; reliability, material availability, operational availability, maintainability, supportability, corrosion, and survivability analysis (new technology and legacy systems).	Legacy O&S costs (e.g., VAMOSC); projected tested reliability and O&S costs of new system components	Estimates are uncertain (especially early in a program) due to projecting new system performance. Detailed sustainment data are not widely shared outside the Service. Externalities can greatly affect actual costs compared to estimates (e.g., changes in operational environments, OPTEMPO, physical and cyber battle damage, quantity, fuel costs, labor costs, healthcare costs, logistical system configuration and investments).

FIGURE B.5 Continued
(figure continues)

Domain	Decision	Supporting Analyses	Analytic Capabilities				Gaps and Challenges
			DoD AWF	Other Resources	Analytic Approaches	Key Data: Available	
Schedule	Schedule estimated and reasonable	Schedule analysis; technical maturity and risk analysis	CE; Contracting; PM	FFRDCs			
Compliance	Program compliant with laws and policies	Legal and policy analysis	All but esp. PM	OGC; FFRDCs			
Contracting	Contracting strategy developed and appropriate	Contracting strategy and analysis	Contracting, FM, PM	Buying commands; FFRDCs	Contract-type determination; Termination liability estimating	Contract execution risks; contractor investments and costs.	
Production	MS C, FRP, and production management	Engineering, manufacturing, T&E, quality, reliability, contracting, audits, and payments	PQM, T&E, Engr., Facilities Engr., IT, Logistics Contracting, Purchasing, Auditing, FM, PM,	Test centers, FFRDCs/UARCs	Numerous specialty-specific approaches	Designs, production approaches, quality, scrap rates, failure rates, reliability	Detailed data rarely available to higher-level oversight organizations
Portfolio	Program satisfies larger portfolio needs		Engr., PM; Leadership	Requirements, FFRDCs, Systems and Buying Commands	Ad hoc	Kill chains; JCAs; operational concepts; Program and system interdependencies and schedules; Current operational systems (commodity, inventory, age, condition, capabilities)	Portfolio analysis and data are largely ad hoc, depending on oversight workforce knowledge and unstructured analysis. Funding flexibility
	Program budget sufficient given larger portfolio interdependences		Engr., FM, PM, Leadership	Requirements, FM, Systems and Buying Commands, FFRDCs	Ad hoc	Program and system interdependencies and schedules	Portfolio analysis and data are largely *ad hoc* by oversight workforce knowledge and unstructured analysis

FIGURE B.5 Continued

al. (2019) to understand the details, but the committee presents this extract to help convey a sense of the depth and range of decision, analysis, and data types involved.

THE DEFENSE ACQUISITION WORKFORCE

The breadth of acquisition activities is also reflected in the range of roles and career fields that constitute the Defense Acquisition Workforce (AWF) (Table B.2). DoD currently identifies and tracks 14 acquisition-specific career fields that cover key roles throughout the acquisition life cycle, which differ from the formal military designators and civilian occupational series designated by the Office of Personnel Management (OPM). The AWF career fields are populated by personnel from various military fields and OPM occupational series. As of the end of FY 2019, there were 180,355 full-time civilian and military personnel in the AWF (DoD 2019).

WORKFORCE IMPROVEMENT AND TRAINING

In 1986, the Packard Commission Report on defense management found that (1) "compared to its industry counterparts, this workforce is undertrained, underpaid, and inexperienced" and (2) "training should be… centrally managed and funded to improve the utilization of teaching faculty, enforce compliance with mandatory training requirements, and to coordinate overall acquisition training policies" (*Inside the Pentagon* 1989).

In response, the Defense Acquisition Workforce Improvement Act (DAWIA),[1] enacted in 1990, requires the Secretary of Defense to establish education and training standards, requirements, and courses for the civilian and military acquisition workforce (Box B.3). Those responsibilities of the Secretary are generally delegated to the Under Secretary of Defense for Acquisition and Sustainment.

As is noted in Box B.3, the Defense Acquisition University (DAU) was established via DAWIA and 10 USC. DAU offers training to military and federal civilian staff and federal contractors, and is headquartered in Fort Belvoir, Virginia. Box B.4 lists the current mission and vison of DAU.

Beyond specific education for acquisition at DAU, AWF members may receive training via the five service academies (U.S. Military Academy, U.S. Naval Academy, U.S. Air Force Academy, U.S. Coast Guard Academy, U.S. Merchant Marine Academy), Air Force Institute of Technology, Naval Postgraduate School, National Defense University, and public and private colleges and universities. Collectively, these institutions offer some programs and courses in data use and analysis, and are adding more to the curriculum.

---

[1] H.R. 5211—101st Congress (1989-1990).

TABLE B.2 Department of Defense Acquisition Workforce Size by Career Field and Component—Civilian and Military (FY2020Q3)

FY 2019 Q4	Army	Navy	Marine Corps	Air Force	4th Estate	Totals	%Total
Auditing	—	—	—	—	3,799	3,799	2.1111%
Business - CE	248	552	35	535	112	1,482	0.8%
Business - FM	1,758	2,446	183	2,304	641	7,332	4.0%
Contracting	8,845	6,401	580	8,505	8,230	32,561	18%
Engineering	9,082	24,501	334	10,263	2,169	46,349	25%
Facilities Engineering	6,848	6,923	38	791	113	14,713	8.1111%
Information Technology	2,021	3,969	235	1,585	1,054	8,864	4.9999%
Life Cycle Logistics	6,873	6,809	668	4,040	3,472	21,862	12%
Production, Quality and Manufacturing	1,370	4,006	25	494	5,573	11,468	6.3%
Program Management	3,533	6,062	782	7,066	1,879	19,322	11%
Property	52	79	27	19	276	426	0.2%
Purchasing	268	375	27	40	460	1,170	0.6666%
S&T Manager	676	534	3	2,916	150	4,279	2.3%
Test and Evaluation	1,941	3,223	125	3,352	371	9,012	4.9999%
Unknown/Other	7	2	1	4	18	32	0.0202020 2%
Totals	43,522	65,882	3,036	41,914	28,317	182,671	
Component %	23.8888%	36.1111%	1.7%	22.9999%	15.5555%		

NOTE: CE = cost estimating; FM = financial management; S&T = science and technology.
SOURCE: Department of Defense, 2019, "Defense Acquisition Workforce Key Information, Overall," briefing, Washington, DC, FY20Q3, June 30, https://www.hci.mil/docs/Workforce_Metrics/FY20Q3/FY20(Q3)OVERALLDefenseAcquisitionWorkforce(DAW)InformationSummary.pptx.

APPENDIX B

> **BOX B.3**
> **Defense Acquisition Workforce Improvement Act (DAWIA) and 10 U.S.C.**
>
> Sec. 1722. Career Development
>
> (a) The Secretary of Defense shall ensure that appropriate **career paths** for civilian and military personnel who wish to pursue careers in acquisition are identified in terms of the education, **training**, experience, and assignments necessary for career progression.
>
> 10 United States Code (USC), Sec 1746.
>
> (a) The Secretary of Defense shall establish and maintain a defense acquisition university structure to provide for—
>   (A) The professional educational development and training of the acquisition workforce
>   (B) Research and analysis of defense acquisition policy issues from an academic perspective
> (b) May employ as many civilians as professors as the Secretary considers necessary

> **BOX B.4**
> **Defense Acquisition University Mission and Vision**
>
> **Mission:** Provide a global learning environment to develop qualified acquisition, requirements and contingency professionals who deliver and sustain effective and affordable warfighting capabilities.
>
> **Vision:** An accomplished and adaptive workforce, giving the warfighter a decisive edge.
>
> SOURCE: Defense Acquisition University, "About DAU," last updated 2020, https://www.dau.edu/about.

REFERENCES

Anton, P.S., M. McKernan, K. Munson, J.G. Kallimani, A. Levedahl, I. Blickstein, J.A. Drezner, and S. Newberry. 2019. *Assessing Department of Defense Use of Data Analytics and Enabling Data Management to Improve Acquisition Outcomes*. RR-3136-OSD. Santa Monica, CA: RAND Corporation. August. https://www.rand.org/pubs/research_reports/RR3136.html.

DoD (Department of Defense). 2019. "Defense Acquisition Workforce Key Information, Overall," briefing, FY20Q3. (June 30.). https://www.hci.mil/docs/Workforce_Metrics/FY20Q3/FY20(Q3)OVERALLDefenseAcquisitionWorkforce(DAW)InformationSummary.pptx.

Inside the Pentagon. 1989. "National Security Review 11 DEFENSE MANAGEMENT." Special Report. July 7. pp. 1-19. http://www.jstor.org/stable/43985715.

C

Data Science Case Studies in Defense Acquisition

The use of data science to support decision making is not new to the acquisition community and has facilitated critical program support and even informed acquisition strategies for decades. However, the advent of increasingly robust data tools, associated visualization techniques and data analytics employed in the context of the burgeoning data science discipline offer the Department of Defense (DoD) an opportunity to institutionalize what once was a sporadic and often ad hoc process into a cohesive, consistent expectation across DoD programs.

This appendix highlights examples where data science has improved DoD acquisition processes. These examples illustrate how data analytics inform the understanding of existing processes and serves to measure the effect of new innovations. Analytics do not provide all the answers because human innovation, leadership, and decision-making are key considerations. What measurement and analysis enable are insights into the effectiveness of a system and how new ideas and approaches impact performance.

CURRENT USES OF DATA IN DOD ACQUISTION PROCESS

Program management in the DoD acquisition community is a data-driven discipline that involves controlling the cost, schedule, and performance of a project or set of projects. Defense acquisition program managers are responsible for achieving development, production, and sustainment objectives of their projects. The following case studies demonstrate how data science has been key to decision making in DoD acquisition programs for decades.

Multiyear Procurement Programs' Defense, the Cost Assessment and Program Evaluation Office, and the Cost Assessment Data Enterprise

Data Science Application: Tracking costs through multi-year programs through common database and analysis platforms.

For decades, the Office of the Secretary of Defense Cost Assessment and Program Evaluation (CAPE) and its predecessors kept a database of contractor costs for DoD acquisition contracts that analysts used to make cost estimates. That database was assembled by individual analysts who manually entered data from individual contractor cost data reports submitted as part of the acquisition contracts. Beginning in 2013, analysts in CAPE set DoD on a path to capture this data automatically. The result of that effort is the Cost Assessment Data Enterprise (CADE).

CADE reduced the data entry burden on both the contractors and CAPE analysts, allowing more time for data analysis. Through CADE, CAPE's goal is "to increase analyst productivity and effectiveness by collecting, organizing and displaying data in an integrated single web-based application...." CADE has made these data readily available in a format that is easily searchable to facilitate better and more consistent cost analysis across the entire DoD.

By using the data in CADE, cost analysts across DoD can more readily calculate the expected single-year cost of a procurement to support effective contract negotiations. This has led to significant savings for DoD. For example, from the FY 2017 Annual Report on Cost Assessment Activities, CAPE completed five analyses supporting multi-year procurements. Three were for aircraft programs, and CAPE estimated the resulting savings ranged from 10 to 13.7 percent (DoD 2018). Two were for ship programs with savings ranging from 5 to 10 percent. The total amount of savings from these five multi-year procurements is approximately $4.5 billion (in then-year dollars).

While DoD cost analysts in CAPE and in the military departments have done analysis to support multi-year procurements for a long time, CADE has made it easier for the analysts and has supported more consistent results across DoD enabling more effective contract negotiations. This enhanced use of data is summarized in Box C.1.

> **BOX C.1
> Multi-year Procurement Programs' Defense,
> the Cost Assessment and Program Evaluation Office, and
> the Cost Assessment Data Enterprise—At a Glance**
>
> Project: Formation of the Cost Assessment Data Enterprise (CADE) via the Office of the Secretary of Defense Cost Assessment and Program Evaluation (CAPE)
>
> Objective: Using a digitized database and accompanying tools, provide high-quality, accessible, and usable historical cost data from previous DoD programs.
>
> Why: The Office of the Secretary of Defense projects costs for programs based on these cost data. The data have long been kept in CAPE, but were recorded manually. CADE was initiated to more easily inform those who do cost analysis—the data are higher quality, provided faster, and are centrally managed.
>
> Impacts: Funding allocation for multi-year appropriation approval, based on the quality, of the data; benchmark for other programs for comparing costs, affordability, budgeting, establishing programs of record.
>
> Challenges: Getting contractors to submit data based on the standards that were established. Finding the talent to curate/manage data.
>
> What worked: Leadership directing compliance. Director, CAPE, and the Under Secretary of Defense for Acquisition and Sustainment both supported the effort. Leadership directing compliance. Dr. Kendall (or someone else) was supportive of this endeavor.
>
> Upskilling: Enabled talented CAPE staff with resources, supported by contractors. CAPE established rotational assignments of government cost analysts to share cost analysis techniques and demonstrate access to CADE.

Weapons System Test and Evaluation

Data Science Application: Supporting statistical modeling and applying common tools and training to enhance testing and evaluation of weapons systems.

A key function of the DoD acquisition workforce is testing and evaluating new systems. The goal of operational test and evaluation is to understand how new or upgraded systems will perform under realistic combat conditions prior to full-rate production and fielding to combat units (DoD 2016). Understanding the capabilities and limitations of systems before they are fielded helps commanders in the field employ the systems effectively and allows for corrective actions before large numbers of systems are produced to help minimize expensive retrofitting.

Test and evaluation is inherently data-driven, as it uses a series of events to gather data to understand the system's effectiveness and suitability. This data supports decisions on design changes, corrective actions, and full-rate production.

Since 2008, the office of the Director of Operational Test and Evaluation (DOT&E) has supported a variety of efforts to upskill the test and evaluation (T&E) workforce in data science. Specific areas of focus include statistical design of experiments (DOT&E 2010), the design and analysis of surveys, design for reliability and statistical reliability assessment (DOT&E 2016), and methodologies for combining information from multiple tests (DOT&E 2016).

These efforts have focused on providing training resources tailored to the T&E community and on developing a community of practice.

- Since 2012, the Institute for Defense Analyses (IDA) has offered a variety of 1- to 3-day on-site courses for the service operational test agencies on topics such as test planning foundations, design of experiments, statistical analysis, and introduction to statistical thinking.
- The Test Science website[1] provides case studies, online tools, free R software packages tailored for T&E professionals (Haman 2020; Morgan-Wall and Khoury 2020), webinars, and training videos.
- The Science of Test (SOT) Research Consortium, sponsored by DOT&E and the Test Resource Management Center, is an academic research consortium supporting faculty experts in experimental design, statistical analysis, reliability, and modeling and simulation from the Naval Postgraduate School, the Air Force Institute of Technology, and six additional universities.
- The SOT Research Consortium provides data science expertise by developing new techniques, producing content for the Test Science website, collaborating with DoD professionals, and training students for the workforce (DOT&E 2016).
- The Defense and Aerospace Test and Evaluation Workshop,[2] sponsored by DOT&E, the National Aeronautics and Space Administration (NASA), the American Statistical Association, and IDA, provides an opportunity for practitioners to interact with researchers, develop collaborations, take short courses, and receive feedback on projects and issues.

---

[1] See https://testscience.org/.
[2] See https://dataworks.testscience.org/.

APPENDIX C   113

These training efforts have had impacts at a variety of levels. In 2007, Congressional concern about body armor testing led to DOT&E working closely with the services to develop test protocols for the ballistic components of combat helmets. Applying statistical design of experiments at a program level allowed DoD to establish a testing and analysis program that identified an initial design for the Enhanced Combat Helmet as providing insufficient protection against small arms penetrations. This analysis led to a key design change in the ballistic shell laminate material of the combat helmet (DOT&E 2016; Hester et al. 2015).

Similarly, a detailed data analysis on the operational evaluation of the Littoral Combat System directly influenced Defense Secretary Hagel's decision to direct the Navy to consider a new frigate.

Within defense acquisition, the T&E community have a long track record of valuing and using data. Training programs are robust, civilian and military career paths support the T&E demand, and a common lexicon is in place. Data are valued and demanded by leadership and decision makers.

Velocity Management: How Data Analysis Helped to Improve the Army's Logistics System

Data Science Application: Tracking maintenance data and replacement parts fulfillment.

In the realm of Army logistics, a classic example of how data science and analytics can enable a major transformation of critical defense systems is the U.S. Army's Velocity Management effort. Starting in 1994, the Army sought to improve its order fulfillment and related processes for parts. As shown in Figure C.1, the key to improving this logistics process was to define and establish performance metrics and associated data to understand how well the system is performing at baseline, diagnosing performance drivers, and informing process improvements. Army logisticians monitor data and metrics and the effects of any changes driving iterative improvements.

The Army improved its data collection systems to better measure data on order submissions, item movement, and delivery locations. These data were a key enabler for the effective use of performance metrics. Figure C.2 illustrates some of the basic elements in the order fulfillment process for Army units in the continental United States; Army units abroad have additional elements (Dumond et al. 2001).

To highlight how data analysis improved Army logistics, consider how Velocity Management examined whether to add or remove an item from the warehouse inventory. Algorithms calculated the tradeoff between the effect of stocking an item and its associated cost. Analysis revealed that

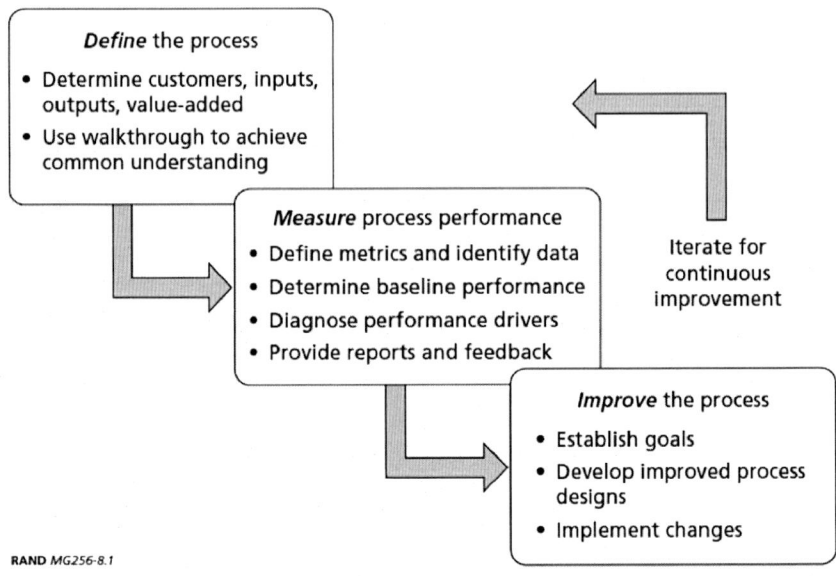

FIGURE C.1 Steps in the Define-Measure-Improve Method. SOURCE: Dumond et al., 2001; Dumond and Eden, 2005. Courtesy of RAND.

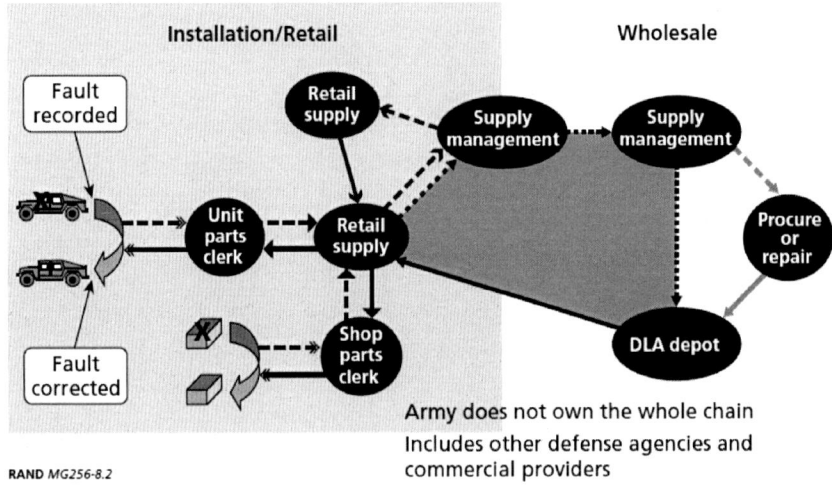

FIGURE C.2 Order fulfillment process for army units in the continental United States. SOURCE: Dumond et al., 2001; Dumond and Eden, 2005. Courtesy of RAND.

stocking additional inexpensive items reduced customer wait time with little investment (Dumond et al. 2001). Many of these low-cost items were small, so stocks remained mobile with little increase in total weight and volume. Other analyses identified items that were no longer in demand and could be removed from storage. In addition, data analytics identified when warehouses and storage containers could be reconfigured to improve space utilization for higher-demand items and generating credit for returning unnecessary items (Dumond et al. 2001).

Figure C.3 illustrates the number of days to fulfill orders from Fort Bragg for repair parts before and after Velocity Management (not counting backordered parts). At the median (the point at which half of the order parts are fulfilled), the delay was reduced from 18 days to 5 days. The delay on three-quarters of ordered parts was reduced from 28 days to 7 days. Thus, military equipment can be put back into the field in about a quarter of the time compared to when Velocity Management was not in place.

These data do not specifically reveal which repairs were accelerated, which equipment are considered critical, and what the associated warfighting effect is of the equipment (i.e., it could be that the most critical equipment still takes a long time while simpler equipment is turned around faster). However, the fact that the speed-up extends into the 95th percentile affects overall operational readiness and availability of the warfighter.

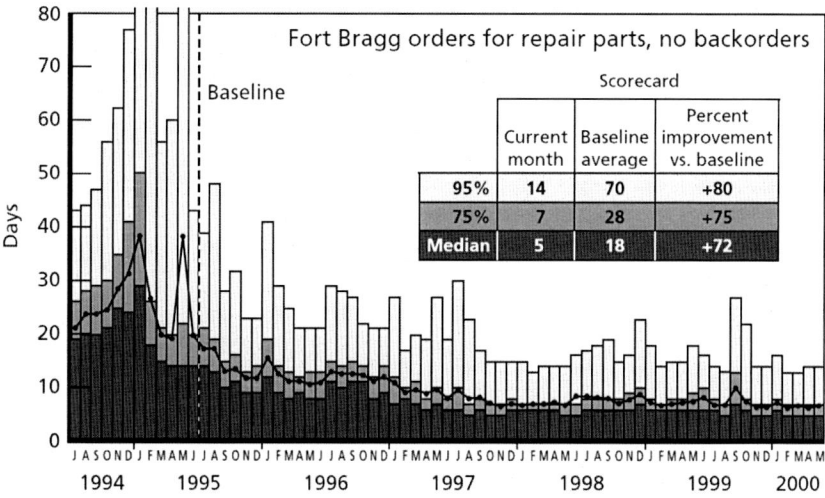

FIGURE C.3 Improvement trends for order fulfillment times before and after velocity management. SOURCE: Dumond and Eden, 2005. Courtesy of RAND.

Dumond and Eden (2005) report that data-driven measurement, metric feedback, simplified performance rules, and new information systems were key elements. These metrics and associated data helped to inform the status of the Army logistics system and demonstrated that other key elements were effective in speeding delivery, including leadership commitment, streamlined reviews, scheduled truck deliveries, higher fulfilment from the main depot, and direct deliveries to customers. A brief summary of this improved use of data for the Army logistics system appears in Box C.2.

Business Analytics Saves Air Force $2 Billion and Counting

Data Science Application: Business analytics to inform acquisition contracting.

Category management is an effort to move from managing purchases and prices individually across thousands of procurement items to managing spend and total cost for entire categories of purchases. In 2014, the Air Force published the Air Force Category Management Concept of Operations (CONOPS), leveraging an industry best practice employed by leading companies (e.g., Walmart, UPS, Kroger). In the category management construct, the Air Force determined common categories of products and services (i.e., heating and air conditioning, airfield paving, working dogs, etc.), assigned Air Force-level responsibility to a category manager with

---

BOX C.2
Velocity Management: How Data Analysis Helped to Improve the Army's Logistics System—At a Glance

Project: Using data science to improve the Army's logistics system.

Objective: Improve the Army's logistics system, specifically its parts fulfillment system and design metrics for performance and diagnosing problems.

Results: Improvements to the Army's logistics system had a direct effect on the availability of military equipment.

- Major decrease in delays in fulfillment of replacement parts
- Military equipment returned to the field in a quarter of the time before Velocity Management was implemented

What worked: Key elements of the program's success were data-driven measurement, metric feedback, simplified performance rules, and new information systems.

expertise in their assigned category and who now manages their categories as a virtual mini-business with its own set of strategies.

The backbone of this Category Management CONOPS is the Business Intelligence Competency Cell (BICC) under the Air Force Installation Contracting Center at Wright-Patterson AFB in Ohio. Its process couples historical spend data and functional data to provide complete "as-is" Air Force benchmarks; engages industry on cost drivers, best practices, benchmarks, and customary commercial practices; and identifies gaps in cost, performance, and implements execution plans to address the gaps.

BICC capitalized on existing commercial data tools. For example, it procured an enterprise license for an industry leading market intelligence service covering more than 900 reports on common commercial goods and services and acquired a license for Tableau, a modern commercial off-the-shelf data visualization software. Using these tools, the BICC team linked existing government data sets such as Federal Procurement Data System-Next Generation, the System for Award Management, and the Contracting Business Intelligence Service, which collects data from Air Force contract writing systems and Government Purchase Card programs to create the Air Force Business Intelligence Tool. This centralized team of 9 people delivered $2 billion in savings in its first 5 years of operation. The BICC team includes data analysts with domain expertise and data engineers. Five of the 9 team members are contracted, which means that BICC could not find the needed skills in the Air Force acquisition workforce so hired the talent by contract. Organizationally, this small team services the entire Air Force on strategic sourcing and category management initiatives. It also supports 78 Air Force Contracting Squadrons, and shares its tools and training with the many others including the Army, Navy, DHS, and GSA.

Data Driven Analysis of Alternatives Restructures C-17 Globemaster II Program

Data Science Application: Analysis of alternatives and cost comparison.

The U.S. Air Force's C-17 Globemaster II is one of the most successful airlift platforms in world and widely flown by the Air Force. In addition to transoceanic flight and supply missions (goods, vehicles, and personnel), the C-17 performs tactical airlift, medical evacuation, and airdrop missions. While widely acknowledged as successful aircraft, the acquisition program to develop and purchase the aircraft was nearly canceled in its infancy due to rising cost and underperformance in the aircraft itself—proving earlier assessments of low development cost and risk inaccurate. This prompted Congress to mandate that DoD conduct an independent assessment of the program that would ultimately be executed by IDA.

IDA based its analysis on the data collected and developed a series of models and simulations to drive the analysis. For instance, to assess the C-17 performance and compare it against alternative models, the IDA developed the Airlift Loading Model (ALM) for loading, the detailed simulation model Mobility Analysis Support System (MASS) for airlift movement and cargo flows, and the Airlift Cycle Analysis Spreadsheet (ACAS) for assessments of cargo and passenger movement. ALM, MASS, and ACAS all served to collate data on aircraft performance and assess it against alternative systems. The IDA team also utilized cost estimates to support its analysis and compliment the performance models' assessment.

Given the effectiveness of individual platforms, the potential of a mixed fleet, and potential changes to Army vehicle development outside of the Air Force's programs purview, the report's final analysis identified and prioritized the options and alternatives for DoD to consider.

While the initial decision following the IDA study was to cap the number of C-17s the Air Force would purchase—40 aircraft—and pursue a mixed fleet that would include 747s, the study prompted a change in the C-17 program. Using data analysis to lay bare how the aircraft compared to alternative options and identifying inefficiencies in their program, the study pushed the C-17 program to restructure itself. This restructuring combined with follow-up studies conducted by the Air Force and DoD with regard to program requirements, led to a decision in 1995 to proceed with the original purchase of 120 C-17s (IDA 2010). A brief summary of the use of performance and cost assessment data for the C-17 program appears in Box C.3.

REFERENCES

DoD (Department of Defense). 2016. *Performance of the Defense Acquisition System: 2016 Annual Report*. Under Secretary of Defense for Acquisition, Technology, and Logistics. October. http://www.dtic.mil/docs/citations/AD1019605.

DoD. 2018. FY 2017 Annual Report on Cost Assessment Activities. February. https://www.cape.osd.mil/files/Reports/CA_AR2017v10Final.pdf.

DOT&E (Director of Operational Test and Evaluation). 2010. "Director, Operational Test and Evaluation (DOT&E) FY 2010 Annual Report." https://www.dote.osd.mil/Publications/Annual-Reports/2016-Annual-Report/.

DOT&E. 2016. "Director, Operational Test and Evaluation (DOT&E) FY 2016 Annual Report." https://www.dote.osd.mil/Publications/Annual-Reports/2016-Annual-Report/.

Dumond et al. 2001. Dumond, J., M.K. Brauner, R. Eden, J.R. Folkeson, K.J. Girardini, D.J. Keyser, E. Peltz, E.M. Pint, and M.Y.D. Wang. 2001. "Velocity Management: The Business Paradigm That Has Transformed U.S. Army Logistics." RAND Corporation. https://www.rand.org/pubs/monograph_reports/MR1108.html.

Dumond, J., and R. Eden. 2005. "Improving Government Processes: From Velocity Management to Presidential. Appointments." RAND Corporation. https://www.rand.org/pubs/reprints/RP1153.html.

Haman, J.T. 2020. September.

BOX C.3
Performance and Independent Cost Assessment for C-17 Globemaster II—At a Glance

Project: Independent assessment of the C-17 Globemaster II program conducted by the Institute for Defense Analyses (IDA).

Objective: Identify alternative systems, establish air lift requirements, estimate effectiveness and total costs of each system, and arrange the data analysis to facilitate decision making.

Results: The report identified and prioritized options and alternatives. The analysis prompted changes in the C-17 program that reversed a previous decision to shrink the number of aircraft the Air Force would purchase and helped the program reach the original purchase quota of 150 aircraft.

Challenges: Disparate data spread across multiple vendors and contractors

What worked: The development and use of multiple databases and models to simulate and evaluate the capabilities and costs for each alternative system allow for comparisons across programs.

Data team: Outsourced to IDA.

Upskilling: Lack of an organic ability to perform cross program data analysis.

Hester, J., T. Johnson, and L. Freeman. 2015. "Managing Risks: Statistically Principled Approaches to Combat Helmet Testing." In Research Notes. Fall. Institute for Defense Analyses, Alexandria, VA.

IDA (Institute for Defense Analysis). 2010. "An Application of Cost-Effectiveness Analysis in a Major Defense Acquisition Program: The Decision by the U.S. Department of Defense to Retain the C-17 Transport Aircraft." https://www.ida.org/-/media/feature/publications/a/an/an-application-of-costeffectiveness-analysis-in-a-major-defense-acquisition-programthe-decision-by/ida-document-d-4218.ashx.

Morgan-Wall, T., and G.A. Khoury. 2020. "Design of Experiments Suite: Generate and Evaluate Optimal Designs [R package skpr version 0.64.2]." Corpus ID: 226069590. https://www.semanticscholar.org.

D

Skills for Data Science Mastery

The National Academies of Sciences, Engineering, and Medicine 2018 report *Data Science for Undergraduates: Opportunities and Options* (The National Academies Press, Washington, DC) emphasized that a critical task in the education of future data scientists is to instill basic data acumen. This requires exposure to key concepts in data science, real-world data and problems that can reinforce the limitations of tools, and ethical considerations that permeate many applications. The following are key concepts involved in developing basic data science acumen.

- *Mathematical foundations.* Key mathematical concepts/skills that would be important for all students in their data science programs and critical for their success in the workforce are the following:
  — Set theory and basic logic,
  — Multivariate thinking via functions and graphical displays,
  — Basic probability theory and randomness,
  — Matrices and basic linear algebra,
  — Networks and graph theory, and
  — Optimization.

  Some data scientists and programs require a deeper understanding of mathematical underpinnings. This might include the following:
  — Partial derivatives (to understand interactions in a model),
  — Advanced linear algebra (i.e., properties of matrices, eigenvalues, decompositions),

— "Big O" notation and analysis of algorithms, and
  — Numerical methods (e.g., approximation and interpolation).

- *Computational foundations.* While it would be ideal for all data scientists to have extensive coursework in computer science, new pathways may be needed to establish appropriate depth in algorithmic thinking and abstraction in a streamlined manner. This might include the following:
  — Basic abstractions,
  — Algorithmic thinking,
  — Programming concepts,
  — Data structures, and
  — Simulations.

- *Statistical foundations.* Important statistical foundations might include the following:
  — Variability, uncertainty, sampling error, and inference;
  — Multivariate thinking;
  — Non-sampling error, design, experiments (e.g., A/B testing), biases, confounding, and causal inference;
  — Exploratory data analysis;
  — Statistical modeling and model assessment; and
  — Simulations and experiments.

- *Data management and curation.* Key data management and curation concepts/skills that would be important for all students in their data science programs and critical for their success in the workforce are the following:
  — Data provenance;
  — Data preparation, especially data cleansing and data transformation;
  — Data management (of a variety of data types);
  — Record retention policies;
  — Data subject privacy;
  — Missing and conflicting data; and
  — Modern databases.

- *Data description and visualization.* Key data description and visualization concepts/skills that would be important for all students in their data science programs and critical for their success in the workforce are the following:
  — Data consistency checking,
  — Exploratory data analysis,

- Grammar of graphics,
- Attractive and sound static and dynamic visualizations, and
- Dashboards.

- *Data modeling and assessment.* Key data modeling and assessment concepts/skills that would be important for all students in their data science programs and critical for their success in the workforce are the following:
  - Machine learning (e.g., supervised, unsupervised, and deep learning),
  - Multivariate modeling and supervised learning,
  - Dimension reduction techniques and unsupervised learning,
  - Deep learning,
  - Model assessment and sensitivity analysis, and
  - Model interpretation (particularly for "black box" models).

- *Workflow and reproducibility.* Key workflow and reproducibility concepts/skills that would be important for all students in their data science programs and critical for their success in the workforce are the following:
  - Workflows and workflow systems,
  - Documentation and code standards,
  - Source code (version) control systems,
  - Reproducible analysis, and
  - Collaboration.

- *Communication and teamwork.* Key communication and teamwork concepts/skills that would be important for all students in their data science programs and critical for their success in the workforce are the following:
  - Ability to understand client needs,
  - Clear and comprehensive reporting,
  - Conflict resolution skills,
  - Well-structured technical writing without jargon, and
  - Effective presentation skills.

- *Domain-specific considerations.* Effective application of data science to a domain requires knowledge of that domain. Grounding data science instruction in substantive contextual examples (which will require the development of judgment and background in those areas) will help ensure that data scientists develop the capacity to pose and answer questions with data. Reinforcing skills and capacities developed in data science courses in the context of a specific domain will help students see the entire data science process.

- *Ethical problem solving.* Key aspects of ethics needed for all data scientists (and for that matter, all educated citizens) include the following:
  — Ethical precepts for data science and codes of conduct,
  — Privacy and confidentiality (both in the spirit and letter of the law),
  — Responsible conduct of research (e.g., human subjects),
  — Ability to identify "junk" science, and
  — Ability to detect algorithmic and human bias.

E

Glossary of Terms, Abbreviations, and Acronyms

A&S	Acquisition and Sustainment
AFIT	Air Force Institute of Technology
AI	artificial intelligence
ASA(ALT)	Assistant Secretary of the Army for Acquisition, Logistics, and Technology
AWF	Acquisition Workforce
BICC	Business Intelligence Competency Cell
CADE	Cost Assessment Data Enterprise
CAPE	Cost Assessment and Program Evaluation
CBO	Congressional Budget Office
CDO	chief data officer
CONOPS	concept of operations
Data acumen	The mastery of key concepts in data science, real-world data and problems that can reinforce the limitations of tools, and ethical considerations that permeate many applications.
Data literacy	Data literacy concepts include data questions; assessment of data sources and quality; margin of error and uncertainty; concerns related to ethics, security, and privacy; basic tools and techniques for visualizing data; interpreting visualizations and results; and best practices for communication.

Data science	The science and technology of extracting value from data and is characterized by the data life cycle—shown here.

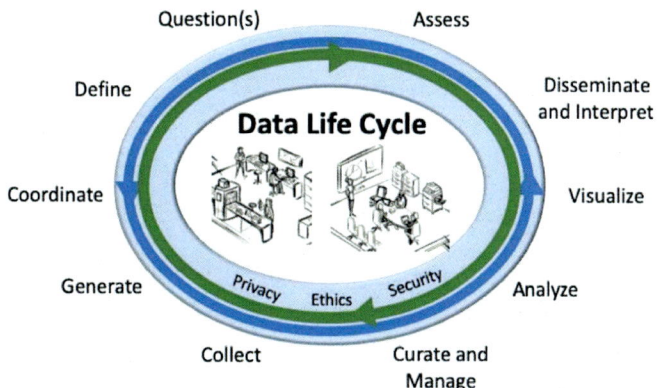

Data strategy	A data strategy for an organization is a guide for how data should be collected, stored, managed, shared, and used within the data life cycle.
DAU	Defense Acquisition University
DAWIA	Defense Acquisition Workforce Improvement Act
Defense acquisition	The conceptualization, initiation, design, development, test, contracting, production, deployment, integrated product support, modification, and disposal of weapons and other systems, supplies, or services (including construction) to satisfy DoD needs, intended for use in, or in support of, military missions. (Department of Defense, 2020, Directive 5135.02. Under Secretary of Defense for Acquisition and Sustainment (USD(A&S)). https://www.esd.whs.mil/Portals/54/Documents/DD/issuances/dodd/513502p.pdf?ver=2020-07-15-133537-05)
DoD	Department of Defense
DON	Department of the Navy
DOT&E	Director of Operational Test and Evaluation
ETL	extract, transform, and load
FTC	Federal Trade Commission
GSA	General Services Administration
IDA	Institute for Defense Analyses
MOOC	Massive Open Online Course
NASEM	National Academies of Sciences, Engineering, and Medicine
NAVAIR	Naval Air System Command
NDU	National Defense University

NPS	Naval Postgraduate School
NRC	National Research Council
OSD	Office of the Secretary of Defense
PQM	production, quality, and manufacturing
R&D	research and development
STEM	science, technology, engineering, and mathematics
T&E	test and evaluation
USD(A&S)	Under Secretary of Defense for Acquisition and Sustainment
USD(AT&L)	Under Secretary of Defense for Acquisition, Technology, and Logistics

F

Committee Member Biographies

WENDY MASIELLO, *Co-Chair*, is president of Wendy Mas Consulting, LLC, and serves on the board of directors for KBR Inc. and EURPAC Service, Inc., and is on the GM Defense LLC Senior Advisory Committee. She also serves as a director on the Procurement Round Table, National Contract Management Association (NCMA), and as an advisor on the Public Spend Forum Council and the Dean's Advisory Council for Texas Tech University's Rawls College of Business. Prior to her July 2017 retirement from the U.S. Air Force, she was director of the Defense Contract Management Agency (2014-2017) where she oversaw a $1.4 billion budget and 12,000 people worldwide in oversight of 20,000 contractors performing 340,000 defense and federal contracts valued at $6 trillion and revised the agency's approach to oversight from one of contractor compliance to performance measurement. During her 36-year career, General Masiello also served as Deputy Assistant Secretary (Contracting), Office of the Assistant Secretary of the Air Force for Acquisition (2011-2014) where she designed and implemented a centralized organization (Air Force Installation Contracting Agency [AFICA]) facilitating contracting for seven major commands following a 30 percent personnel cut. AFICA became the model for the Air Force-wide Installation Mission Support Center. As program executive officer for the Air Forces' $65 billion service acquisition portfolio (2007-2011), General Masiello initiated the category management concept within the Air Force and set a standard for service contract planning, execution, and oversight within the Department of Defense (DoD). General Masiello's medals and commendations include the Defense Superior Service Medal, Distinguished Service Medal, and the Bronze Star. She earned her bachelor of business administration degree from Texas Tech University, a master of science degree in logistics management from the Air Force Institute of Technology, a master of science degree in national resource strategy from the Industrial College of the Armed Forces, Fort Lesley J. McNair, Washington, D.C., and is a graduate of Harvard Kennedy School's Senior Managers in Government. General Masiello is a 2017 Distinguished Alum of Texas Tech University, was twice (2015 and 2016) named among Executive Mosaic's Wash 100, the 2014 Greater Washington Government Contractor "Public Sector Partner of the Year," and recognized by Federal Computer Week as one of "The 2011 Federal 100." She is an NCMA certified professional contract manager as well as an NCMA fellow.

REBECCA NUGENT, *Co-Chair*, is the Stephen E. and Joyce Fienberg Professor of Statistics & Data Science, the associate department head and co-director of Undergraduate Studies for the Carnegie Mellon University Statistics and Data Science Department, and an affiliated faculty member of the Block Center for Technology and Society. She received her Ph.D. in statistics from the University of Washington in 2006. Prior to that, she received her B.A. in mathematics, statistics, and Spanish from Rice University and her M.S. in statistics from Stanford University. She has won several national and university teaching awards including the American Statistical Association (ASA) Waller Award for Innovation in Statistics Education and serves as one of the co-editors of the Springer Texts in Statistics. She recently served on the National Academies of Sciences, Engineering, and Medicine's study on Envisioning the Data Science Discipline: The Undergraduate Perspective. She is the founding director of the Statistics & Data Science Corporate Capstone program, an experiential learning initiative that matches groups of faculty and students with data science problems in industry, nonprofits, and government organizations. She has worked extensively in clustering and classification methodology with an emphasis on high-dimensional, big data problems and record linkage applications. Her current research focus is the development and deployment of low-barrier data analysis platforms that allow for adaptive instruction and the study of data science as a science.

PHILIP S. ANTON is the chief scientist of the Acquisition Innovation and Research Center (AIRC) at the Stevens Institute of Technology. He assesses the practical needs of DoD, helps to envision and develop innovative acquisition research in the AIRC, and ensures the transition and application of AIRC results in DoD acquisition policies, processes, reports, and workforce development. From 1998 to 2021, Dr. Anton was a senior information scientist at the RAND Corporation, where he conducted research on acquisition and sustainment policy, cybersecurity, emerging technologies, technology foresight, process performance measurement and efficiency, aeronautics test infrastructure, and military modeling and simulation. From 2011 to 2016, Dr. Anton served two Pentagon tours as the deputy director for Acquisition Policy Analysis, reporting directly to the Under Secretary of Defense for Acquisition, Technology, and Logistics. Dr. Anton led his center in conducting strategic initiatives to measure and improve the performance of DoD's policies, workforce, and institutions, crafting affordability policy and bringing new analytic insights into the performance of acquisition and sustainment policies, processes, and tradecraft. For these contributions Dr. Anton received the Secretary of Defense Medal for Outstanding Public Service in 2017. From 2004 to 2011, he was the director of the Acquisition and Technology Policy Center in RAND's National Security Research Division. From 1992 to 1998, Dr. Anton managed and conducted artificial intelligence research at the MITRE Corporation. Before graduate school, he worked at Hughes Aircraft and held intern positions at TRW, Rockwell, Aerojet ElectroSystems, and Scott Environmental Technology. Dr. Anton earned his Ph.D. and M.S. in information and computer science from the University of California, Irvine, specializing in computational neuroscience and artificial intelligence. He holds a B.S. in engineering from University of California, Los Angeles (UCLA). He holds a B.S. in engineering from UCLA, specializing in computer engineering.

TRILCE ESTRADA is an assistant professor in the Department of Computer Science at the University of New Mexico. She earned her Ph.D. in 2012 from University of Delaware. Her research interests include self-managed distributed systems, big data analysis, crowd sourcing, and machine learning. Her overarching research goal is to solve computationally intensive and data intensive problems in science, health, and education, especially in scenarios where resources and trained professionals are scarce. She is also actively involved in improving participation of women in computing-related fields.

STEPHEN R. FORREST is the Peter A. Franken Distinguished University Professor of Engineering and Paul G. Goebel Professor of Electrical Engineering and Computer Science, Physics, and Materials Science and Engineering. In 1985, Prof. Forrest joined University of Southern California and, in 1992, moved to Princeton University. In 2006, he rejoined the University of Michigan as vice president for research, where he is the Peter A. Franken Distinguished University Professor. A fellow of the the American Physical Society, the Institute of Electrical and Electronics Engineers, and the Optical Society, and a member of the National Academy of Engineering, the National Academy of Sciences, and the National Academy of Inventors, he has received numerous awards and medals for his invention of phosphorescent organic light-emitting diodes (OLEDs), innovations in OLEDs, organic thin films, and advances in photodetectors for optical communications. Prof. Forrest has authored ~600 papers and has 339 patents. He is co-founder or founding participant in several companies, including Sensors Unlimited, Epitaxx, NanoFlex Power, Universal Display, and Apogee Photonics, and is on the board of directors of Applied Materials. He is past chairman of the board of the University Musical Society and served as chairman of the board of Ann Arbor SPARK, the regional economic development organization and is now on its board of directors. He has served on the board of governors of the Technion—Israel Institute of Technology where he is a Distinguished Visiting Professor of Electrical Engineering. Currently, Prof. Forrest serves as lead editor of *Physical Review Applied* and recently joined the Air Force Studies Board of the National Academies. He received his B.A. in physics from the University of California, Berkeley, and his M.S. and Ph.D. in physics from the University of Michigan.

CHRISTINE H. FOX became the assistant director for policy and analysis of the Johns Hopkins University Applied Physics Laboratory (APL) in 2014. As the nation's largest University Affiliated Research Center, APL performs research and development on behalf of DoD, the intelligence community, the National Aeronautics and Space Administration (NASA), and other federal agencies. The laboratory has more than 5,000 staff members who are making critical contributions to a wide variety of nationally and globally significant technical and scientific challenges. Previously, she served as Acting Deputy Secretary of Defense between December 2013 and May 2014. Until August 2013, Ms. Fox served as the director, Cost Assessment and Program Evaluation in the Office of the Secretary of Defense. She was appointed to that position in November 2009. A presidential appointee confirmed by the U.S. Senate, Ms. Fox served as the principal staff assistant to the Secretary of Defense for analyzing and evaluating plans, programs, and budgets in relation to U.S. defense objectives and resource constraints. Ms. Fox possesses three decades of experience as an analyst and research manager focusing on defense issues, with a special emphasis on operations. She formerly served as the president of the Center for Naval Analyses (CNA), a federally funded research and development center, and as the scientific analyst to the Chief of Naval Operations. Prior to her appointment as president of CNA, Ms. Fox was the vice-president and director of CNA's Operations Evaluation Group, responsible for approximately 85 field representatives focused on helping operational commanders execute their missions. She oversaw CNA's analysis of real-world operations, including the operations in Bosnia and Kosovo in the 1990s, operations in Afghanistan immediately following the September 11 attacks, and the operation in Iraq in early 2003.

She served as a member of NASA's Return to Flight Task Group, chartered by the NASA Administrator to certify the recommendations made by the Columbia Accident Investigation Board. She was also a member of the advisory board of the Applied Physics Laboratory, University of Washington, from 2007 until 2009. Hon. Fox earned a bachelor of science degree in mathematics and a master of science degree in applied mathematics from George Mason University.

MELVIN GREER is chief data scientist, Americas, Intel Corporation. He is responsible for building Intel's data science platform through graph analytics, machine learning, and cognitive computing to accelerate transformation of data into a strategic asset for public sector and commercial enterprises. His systems and software engineering experience has resulted in patented inventions in cloud computing, synthetic biology, and Internet of Things bio-sensors for edge analytics. He functions as a principal investigator (PI) in advanced research studies, including nanotechnology, additive manufacturing, and gamification. He significantly advances the body of knowledge in basic research and critical, highly advanced engineering and scientific disciplines. Mr. Greer is a member of the American Association for the Advancement of Science (AAAS) and the U.S. National Academies of Sciences, Engineering, and Medicine's Government-University-Industry Research Roundtable. Mr. Greer received his bachelor of science degree in computer information systems and technology and his master of science in information systems from American University, Washington, D.C. He also completed the Executive Leadership Program at the Cornell University, Johnson Graduate School and the Entrepreneurial Finance program at Massachusetts Institute of Technology (MIT) Sloan School of Management.

CHARLES ISBELL received his bachelor's in information and computer science from Georgia Tech, and his M.S. and Ph.D. at MIT's Artificial Intelligence Lab. Upon graduation, he worked at AT&T Labs/Research until 2002, when he returned to Georgia Tech to join the faculty as an assistant professor. He has served many roles since returning and is now The John P. Imlay Dean of the College of Computing. Dr. Isbell's research interests are varied, but the unifying theme of his work has been using machine learning to build autonomous agents who engage directly with humans. His work has been featured in the popular press, congressional testimony, and in several technical collections. In parallel, Dr. Isbell has also pursued reform in computing education. He was a chief architect of Threads, Georgia Tech's structuring principle for computing curricula. Dr. Isbell was also an architect for Georgia Tech's first-of-its-kind MOOC-supported M.S. in computer science. Both efforts have received international attention, and have been presented in the academic and popular press. In all his roles, he has continued to focus on issues of broadening participation in computing, and is the founding Executive Director for the Constellations Center for Equity in Computing. He is an Association for the Advancement of Artificial Intelligence fellow and a fellow of the Association for Computing Machinery (ACM). Appropriately, his citation for ACM Fellow reads "for contributions to interactive machine learning; and for contributions to increasing access and diversity in computing."

PETER LEVINE is senior fellow at the Institute for Defense Analyses in Alexandria, Va. Previously, he served as principal assistant and advisor to the Secretary and Deputy Secretary of Defense on readiness; National Guard and Reserve component affairs; health affairs; training; and personnel requirements and management, including equal opportunity, morale, welfare, recreation, and quality of life. Prior to assuming this role, Mr. Levine served from May 2015 to April 2016 as the deputy chief management officer (DCMO) of the Department of Defense. As DCMO, he served as the senior advisor to the Secretary of Defense and the Deputy Secretary of Defense on business transformation and led the department's efforts to streamline business processes and achieve greater efficiencies in management, headquarters, and overhead functions. Prior to his appointment as DCMO, Mr. Levine served on the staff of the Senate Armed Services Committee from August 1996 to February 2015, including 2 years as staff director, eight years as general counsel, and eight years as minority counsel. Throughout this period, Mr. Levine was responsible for providing legal advice on legislation and nominations, and advised members of the committee on acquisition policy, civilian personnel policy, and defense management issues affecting DoD. Mr. Levine played an important role in the enactment of the Military Commissions Act of 2009, the Weapon Systems Acquisition Reform Act of 2009, the Acquisition Improvement and Accountability Act of 2007, the Detainee Treatment Act of 2005, and numerous defense authorization acts. Mr. Levine served as counsel to Senator Carl Levin of Michigan from 1995 to 1996 and as counsel to the Subcommittee on Oversight of Governmental Management of the Senate Committee on Governmental Affairs from 1987 to 1994. In this capacity, Mr. Levine played a key role in the enactment of the Lobbying Disclosure Act of 1995, the Federal Acquisition Streamlining Act of 1994, and the Whistleblower Protection Act of 1989. Mr. Levine was an associate at the law firm Crowell and Moring from 1983 to 1987. He received a bachelor of arts degree summa cum laude from Harvard College and a juris doctor degree magna cum laude from Harvard Law School.

ANN F. McKENNA is the vice dean of strategic advancement for the Ira A. Fulton Schools of Engineering at Arizona State University (ASU) and is a professor of engineering in the Polytechnic School, one of the six Fulton Schools. Dr. McKenna's research focuses on entrepreneurial thinking in the context of engineering faculty mentorship and curricular innovations, design teaching and learning, the role of adaptive expertise in design and innovation, and the impact and diffusion of education innovations. She was named one of nine 2019 American Society for Engineering Education (ASEE) fellows for demonstrating outstanding contributions to engineering education. Dr. McKenna has been an ASEE member since 1996. Dr. McKenna is PI on the National Science Foundation (NSF)-funded ASU Revolutionizing Engineering Departments project that focuses on instilling an additive innovation and risk-taking mindset among faculty to transform engineering teaching practices. She is also PI on the Kern Family Foundation project that is conceptualizing and implementing a national-focused effort on applying an entrepreneurial mindset approach to faculty mentorship. She was a co-investigator and instructor for the first I-Corps for Learning project, which fosters an entrepreneurial mindset in the education community to design and implement novel and effective teaching strategies, technologies, and curriculum materials. Dr. McKenna has twice been the recipient of the ASEE best overall paper award (1998 and 2011), as well as the recipient of the outstanding paper award from the IEEE/ASEE Frontiers in Education conference (1997). Her work in the area of design education has been nationally recognized by receiving the best paper award for three consecutive years, 2009, 2010, and 2011 in the Design in Engineering Education Division of ASEE. She has also received the best research paper (2018) and best teaching paper (2017) in the Entrepreneurship and Engineering Innovation Division of ASEE. Dr. McKenna works across the disciplinary lines of engineering, education, and design and has published in diverse disciplinary venues including *Science*, *Journal of Engineering Education*, *IEEE Computer*, *ASME Journal of Mechanical Design*, and *Teaching in Higher Education*. Dr. McKenna recently served as a senior associate editor for the *Journal of Engineering Education* (2012-2015), the leading research journal in the field of engineering education. She served a two-year term (2011-2013) as a director of the Educational Research and Methods Division of ASEE. She was a member of the advisory board for the National Academy of Engineering Frontiers of Engineering Education Symposium (2011-2013), as well as a panel member for Canada's Natural Sciences and Engineering Research Council's Chairs in Design Engineering program (2011-2014). Prior to joining ASU, she served as a program director at the NSF in the Division of Undergraduate Education and was the director of education improvement in the McCormick School of Engineering at Northwestern University. Dr. McKenna received her bachelor's and master's degrees in mechanical engineering from Drexel University and doctorate from the University of California, Berkeley.

ALYSON G. WILSON is a professor in the Department of Statistics and principal investigator for the Laboratory for Analytic Sciences at North Carolina State University (NCSU). She is a fellow of the ASA and the AAAS. Her research interests include statistical reliability, Bayesian methods, and the application of statistics to problems in defense and national security. Prior to joining NCSU, Dr. Wilson was a research staff member at the Institute for Defense Analyses' Science and Technology Policy Institute (2011-2013), an associate professor in the Department of Statistics at Iowa State University (2008-2011), a technical staff member in the Statistical Sciences Group at Los Alamos National Laboratory (1999-2008), and a senior statistician and operations research analyst with Cowboy Programming Resources (1995-1999). Dr. Wilson received her Ph.D. in statistics from Duke University, her M.S. in statistics from Carnegie Mellon University, and her B.A. in mathematical sciences from Rice University.

JUN ZHUANG is a professor and director of the Decision, Risk and Data Laboratory, Department of Industrial and Systems Engineering, School of Engineering and Applied Sciences, at the University at Buffalo. Dr. Zhuang received a Ph.D. in industrial engineering in 2008 from the University of Wisconsin-Madison. Dr. Zhuang's long-term research goal is to integrate operations research, big data analytics, game theory, and decision analysis to improve mitigation, preparedness, response, and recovery for natural and manmade disasters. Other areas of interest include applications to health care, sports, transportation, supply chain management, sustainability, and architecture. Dr. Zhuang's research has been supported by NSF, the U.S. Department of Homeland Security, the U.S. Department of Energy, the U.S. Air Force Office of Scientific Research (AFOSR), and the National Fire Protection Association. Dr. Zhuang is a fellow of the 2011 U.S. Air Force Summer Faculty Fellowship Program, sponsored by the AFOSR, and a fellow of the 2009-2010 Next Generation of Hazards and Disasters Researchers Program, sponsored by NSF. Dr. Zhuang has published more than 100 peer-reviewed journal articles in *Operations Research*, *IISE Transactions*, *Risk Analysis*, *Decision Analysis*, and *European Journal of Operational Research*, among others. His research and educational activities have been highlighted in *The New York Times*, *The Wall Street Journal*, *Spark CBC Radio*, *Metro*, *The Washington Post*, *USA Today*, *Stanford GSB News*, *NSF Discovery*, *Science Daily*, *Industrial Engineer*, *The Council on Undergraduate Research Quarterly*, and *The Pre-Engineering Times*, among others. Dr. Zhuang is dedicated to mentoring high school, undergraduate, and graduate students in research.